Plumber's Quick-Reference Manual

Other McGraw-Hill Books of Interest

Plumber's Quick-Reference Manual

R. Dodge Woodson

McGraw-Hill

New York   San Francisco   Washington, D.C.   Auckland   Bogotá
Caracas   Lisbon   London   Madrid   Mexico City   Milan
Montreal   New Delhi   San Juan   Singapore
Sydney   Tokyo   Toronto

Library of Congress Cataloging-in-Publication Data

Woodson, R. Dodge (Roger Dodge), date.
    Plumber's quick-reference manual / R. Dodge Woodson.
        p.   cm.
    Includes index.
    ISBN 0-07-071799-0
    1. Plumbing—Handbooks, manuals, etc.   I. Title.
TH6125.W563   1995
696′.1—dc20   95-41567
                                          CIP

McGraw-Hill

*A Division of The **McGraw·Hill** Companies*

1 2 3 4 5 6 7 8 9 0   DOC/DOC   9 0 0 9 8 7 6 5

ISBN 0-07-071799-0

The sponsoring editor for this book was Larry S. Hager, the editing supervisor was Fred Dahl, and the production supervisor was Pamela A. Pelton. It was set in Century Schoolbook by Inkwell Publishing Services.

Printed and bound by R. R. Donnelley & Sons Company

McGraw-Hill books are available at special quantity discounts to use as premiums and sales promotions, or for use in corporate training programs. For more information, please write to the Director of Special Sales, McGraw-Hill, 11 West 19th Street, New York, NY 10011. Or contact your local bookstore.

 This book is printed on recycled, acid-free paper containing a minimum of 50% recycled de-inked fiber.

*This book is dedicated to
my fantastic family:
Kimberley, Afton, and Adam.*

Contents

2. Reference Tables 18

3. Sizing Pipe for Sanitary Drains   **55**

4. Sizing Vent Pipes   **61**

Acknowledgments

I would like to thank my parents, Maralou and Woody, for their help and support throughout my life. Not only have they been great parents; they are good friends.

Dawn Gearld-Hall deserves a lot of credit for this book. She created many of the illustrations used in these pages, and for that, I thank her.

R. DODGE WOODSON

Acknowledgments

Introduction

Have you often wished for a fast, on-the-job reference that could make your life easier? Is so, this is it. Unlike books that bog you down with a lot of words that camouflage the pertinent information you need in a hurry, this book puts plenty of useful data right at your fingertips. You won't have to wade through countless pages of text to figure out how much water a 3-in. stack requires to fill it for testing. No, the fast-action formulas, tables, and illustrations in this book make your on-the-spot calculations fast and easy.

Long-reading books are fine when you are in the mood to improve your knowledge in a leisurely way, but they can create a lot of frustration when you have an inspector coming in 20 minutes and you need an answer to a code question right now. The layout of this book is simple, direct, and user-friendly. Sure, there's text in it, but hundreds of examples give you instant access to those important facts you need. Putting this book in your truck is one of the best things you can do for yourself and your career.

How good is the information here? Will you really use it? Well, the data have been compiled by a seasoned master plumber who has spent more than 20 years in the field. With this kind of experience behind the making of this book, you can bet it is filled with facts, figures, and formulas that you need to know.

When was the last time you used your algebra skills? How long has it been since you tested your trigonometry knowledge? Does finding the area of a circle confuse you? Have you struggled with written instructions on how to perform plumbing math? Well, you're going to love the first couple of chapters in this book. They are filled with easy-to-understand math principles and conversion charts. That's right, you won't have to do a lot of the math yourself. You can simply run your finger down the tables to find your answers. What could be easier?

How much do you know about the code requirements for plumbing fixtures for handicapped persons? Would you know what to do if you were asked to work on a sterilizer in a dental office? Code requirements pertaining to fixtures are extensive, and they can be confusing. That's not a problem when you have this book with you. Chapter 11 is a comprehensive collection of the rules and regulations that apply to plumbing fixtures. And there are many, many other sections of this book that give equal treatment to other phases of plumbing.

Does sizing pipe give you trouble? Would you like some easy way to figure out what size water pipe is required on your next job? The answers are in Chap. 5. Other chapters show you how to size drains and vents.

Take a moment to look over the table of contents for this book. Then flip through the pages and notice the hundreds of examples. There's even information on gas piping in Chap. 16. Really now, have you ever seen a more comprehensive book that is so easy to use and understand? Go ahead and look through it a while. I'm sure you'll agree with me when I say that this is one of the handiest, best books you can buy to improve your plumbing career.

Three major plumbing codes are in use in the United States. The different codes share many characteristics, but they are not all the same. As we discuss code issues in this book, I refer to certain zones. The three zones are referenced as one, two, and three. This numbering system allows you to determine if the code issues being discussed are tied direct-ly to the plumbing code most likely to be in use in your lo-cation. Please refer to the following lists to pinpoint the zone that includes the state where you work. Keep in mind that local codes are subject to modification, so the exact code re-quirements in your jurisdiction may not be identical to the representations presented in this text.

States in Zone One

Arizona
California
Colorado
Idaho
Indiana
Iowa
Kansas
Minnesota
Montana
Nebraska
Nevada
New Mexico
North Dakota
Oregon
South Dakota
Utah
Washington
Wyoming
Parts of
   Texas

States in Zone Two

Alabama
Arkansas
Florida
Georgia
Louisiana
Mississippi
North Carolina
South Carolina
Tennessee
Parts of:
   Delaware
   Maryland
   Oklahoma
   Texas
   West Virginia

States in Zone Three

Connecticut
Illinois
Kentucky
Massachusetts
Michigan
Missouri
New Hampshire
New Jersey
New York
Ohio
Pennsylvania
Rhode Island
Vermont
Virginia
Parts of:
    Delaware
    Maryland
    Maine
    Oklahoma
    West Virginia

1

Plumbing Math

Plumbing math is not a lot of fun to do, but there are times when it is a necessary part of the trade. I'm not talking about simply using a tape measure and adding up a few numbers. The complicated stuff comes when you have to determine the volume of water contained in a vertical stack or something along those lines. Most of us had some form of advanced math in school, but as plumbers we don't find a lot of occasions to use what we learned. For this reason, we tend to forget formulas and procedures that we do need occasionally.

If you were never good at math, you're not alone. Many plumbers don't possess perfect math skills. That's okay. With the help of some basic principles and formulas, you can get by just fine. I'm not going to lecture you in geometry or algebra. No, instead, I'm going to provide you with some easy to understand illustrations that will take the guesswork out of your plumbing math.

Do you remember how to find the square root of a number? How long has it been since you cubed a number or worked out a cube root? If you're like most of us, you rely on calculators, computers, and tables to do most of your complex math problems. There is certainly nothing wrong with this approach. The key is finding the correct answers in the quickest way possible. Look over the tables and formulas that follow, and I think you will be pleased by the amount of time they can save you in the field.

USEFUL FORMULAS

Circumference of a circle = π × diameter or 3.1416 x diameter

Diameter of a circle = circumference × 0.31831

Area of a square = length × width

Area of a rectangle = length × width

Area of a parallelogram = base × perpendicular height

Area of a triangle = 1/2 base × perpendicular height

Area of a circle = π radius squared or diameter squared × 0.7854

Area of an ellipse = length × width × 0.7854

Volume of a cube or rectangular prism = length × width × height

Volume of a triangular prism = area of triangle × length

Volume of a sphere = diameter cubed × 0.5236 (diameter × diameter × diameter × 0.5236)

Volume of a cone = π × radius squared × 1/3 height

Volume of a cylinder = π × radius squared × height

Length of one side of a square × 1.128 = the diameter of an equal circle

Doubling the diameter of a pipe or cylinder increases its capacity 4 times

The pressure (in lb/sq in.) of a column of water = the height of the column (in feet) × 0.434

The capacity of a pipe or tank (in U.S. gallons) = the diameter squared (in inches) × the length (in inches) × 0.0034

1 gal water = 8⅓ lb = 231 cu in.

1 cu ft water = 62½ lb = 7½ gal

ABBREVIATIONS

A or a	Area, acre
AWG	American Wire Gauge
B or b	Breadth
bbl	Barrels
bhp	Brake horsepower
BM	Board measure
Btu	British thermal units
BWG	Birmingham Wire Gauge
B & S	Brown and Sharpe Wire Gauge (American Wire Gauge)
C of g	Center of gravity
cond	Condensing
cu	Cubic
cyl	Cylinder
D or d	Depth, diameter
dr	Dram
evap	Evaporation
F	Coefficient of friction; Fahrenheit
F or f	Force, factor of safety
ft (or ′)	Foot
ft lb	Foot pound
fur	Furlong
gal	Gallon
gi	Gill
ha	Hectare
H or h	Height, head of water
HP	Horsepower
IHP	Indicated horsepower
in (or ″)	Inch
L or l	Length
lb	Pound
lb/sq in.	Pounds per square inch
mi	Mile
o.d.	Outside diameter (pipes)
oz	Ounces
pt	Pint
P or p	Pressure, load
psi	Pounds per square inch
R or r	Radius
rpm	Revolutions per minute
sq ft	Square foot
sq in.	Square inch
sq yd	Square yard
T or t	Thickness, temperature
temp	Temperature
V or v	Velocity
vol	Volume
W or w	Weight
W. I.	Wrought iron

PLUMBING DEFINITIONS

Abstract number	A number that does not refer to any particular object.
Acute triangle	A triangle in which each of the three angles is less than 90°.
Altitude of a triangle	A line drawn perpendicular to the base from the angle opposite.
Angle	The difference in direction of two proceeding from the same point called the *vertex.*
Area	The surface included within the lines that bound a figure.
Arithmetic	The science of numbers and the art of computation.
Base of a triangle	The side on which a triangle is supposed to stand.
Board measure	A unit for measuring lumber. A BM is the volume of a board 12 in. wide, 1 ft long and 1 in. thick.
Circle	A plane figure bounded by a curved line, called the *circumference,* every point of which is equally distant from a point within, called the *center.*
Complex fraction	A fraction the numerator or denominator of which is itself a fraction.
Cone	A body having a circular base and a convex surface that tapers uniformly to the vertex.
Cubic measure	A measure of volume involving three dimensions—length, width, and thickness (depth).
Cylinder	A body bounded by a uniformly curved surface, its ends being equal and parallel circles.
Decimal scale	A scale in which the order of progression is uniformly 10.
Diameter of a circle	A line that passes through the center of a circle and is terminated at both ends by the circumference.
Diameter of a sphere	A straight line that passes through the center of a sphere and is terminated at both ends by the sphere's surface.
Equilateral triangle	A triangle with equal sides.
Even number	A number that can be exactly divided by two.
Exact divisor of a number	A whole number that will divide a number without leaving a remainder.

PLUMBING DEFINITIONS *(cont.)*

Factors	Two or more quantities that, when multiplied, produce a given quantity.
Factors of a number	Numbers that, when multiplied, make that specific number.
Fraction	A number that expresses part of a whole thing or quantity.
Geometry	The branch of mathematics that treats space and its relations.
Greatest common divisor	The greatest number that will exactly divide two or more numbers.
Hypotenuse of a right triangle	The side opposite the right angle.
Improper fraction	A fraction in which the numerator equals or exceeds the denominator.
Isosceles triangle	A triangle with two equal sides.
Least common multiple	The lowest number that is exactly divisible by two or more numbers.
Measure	The extent, quantity, capacity, volume, or dimensions ascertained by some fixed standard.
Mensuration	The process of measuring.
Number	A unit or collection of units.
Odd number	A number that cannot be exactly divided by two.
Parallelogram	A quadrilateral with opposite sides that are parallel and equal.
Percentage	The rate per hundred.
Perimeter	The distance around a figure; the sum of the sides of a figure.
Perpendicular of a right triangle	The side that forms a right angle to the base.
Proper fraction	A fraction in which the numerator is less than the denominator.
Pyramid	A body having for its base a polygon and for its other sides, or facets, three or more triangles that terminate in a common point called the *vertex*.
Quantity	An aspect that can be increased, diminished, or measured.
Radius of a circle	A line extending from the center of a circle to any point on the surface.

PLUMBING DEFINITIONS *(cont.)*

Rectangle	A parallelogram in which all angles are right angles.
Right triangle	A triangle with a right angle (90°).
Scale	The order of progression on which any system of notation is founded.
Scalene triangle	A triangle all sides of which are unequal.
Sphere	A solid body bounded by a uniformly curved surface, all the points of which are equally distant from a point within that is called the *center.*
Square	A rectangle all sides of which are equal.
Trapezoid	A quadrilateral that has two parallel and two oblique sides.
Triangle	A plane figure bounded by three sides and having three angles.
Uniform scale	A scale in which the order or progression is the same throughout the entire succession of units.
Unit	A single thing or a definite quantity.
Varying scale	A scale in which the order of progression is not the same throughout the entire succession of units.

TRIGONOMETRY

Sine	$\sin = \dfrac{\text{side opposite}}{\text{hypotenuse}}$
Cosine	$\cos = \dfrac{\text{side adjacent}}{\text{hypotenuse}}$
Tangent	$\tan = \dfrac{\text{side opposite}}{\text{side adjacent}}$
Cosecant	$\csc = \dfrac{\text{hypotenuse}}{\text{side opposite}}$
Secant	$\sec = \dfrac{\text{hypotenuse}}{\text{side adjacent}}$
Cotangent	$\cot = \dfrac{\text{side adjacent}}{\text{side opposite}}$

ESTIMATING CUBIC YARDS OF CONCRETE FOR SLABS, WALKS, AND DRIVES

Slab Thickness (inches)	Slab Area (square feet)				
1	10	50	100	300	500
2	0.1	0.3	0.6	1.9	3.1
3	0.1	0.5	0.9	2.8	4.7
4	0.1	0.6	1.2	3.7	6.2
5	0.2	0.7	1.5	4.7	7.2
6	0.2	0.9	1.9	5.6	9.3

ATMOSPHERIC PRESSURE PER SQUARE INCH

Barometer (in. Hg)	Pressure (lb/sq in.)
28.00	13.75
28.25	13.88
28.50	14.00
28.75	14.12
29.00	14.24
29.25	14.37
29.50	14.49
29.75	14.61
29.921	14.696
30.00	14.74
30.25	14.86
30.50	14.98
30.75	15.10
31.00	15.23

Rule: Barometer in in. Hg $\times$ 0.49116 = lb/sq in.

POLYGONS

Pentagon	5 sides
Hexagon	6 sides
Heptagon	7 sides
Octagon	8 sides
Nonagon	9 sides
Decagon	10 sides

Polygons are numerous figures having more than four sides. They are named according to the number of sides. To find the area of a polygon: Multiply the sum of the sides (perimeter of the polygon) by the perpendicular dropped from its center to one of its sides, and half the product will be the area. This rule applies to all regular polygons.

INCHES CONVERTED TO DECIMALS OF FEET

Inches	Decimal of a Foot
1/8	0.01042
1/4	0.02083
3/8	0.03125
1/2	0.04167
5/8	0.05208
3/4	0.06250
7/8	0.07291
1	0.08333
1⅛	0.09375
1¼	0.10417
1⅜	0.11458
1½	0.12500
1⅝	0.13542
1¾	0.14583
1⅞	0.15625
2	0.16666
2⅛	0.17708
2¼	0.18750
2⅜	0.19792
2½	0.20833
2⅝	0.21875
2¾	0.22917
2⅞	0.23959
3	0.25000

Note: To change inches to decimals of a foot, divide by 12. To change decimals of a foot to inches, multiply by 12.

DECIMAL EQUIVALENTS OF FRACTIONS OF AN INCH

Inches	Decimal of an Inch
1/64	0.015625
1/32	0.03125
3/64	0.046875
1/16	0.0625
5/64	0.078125
3/32	0.09375
7/64	0.109375
1/8	0.125
9/64	0.140625
5/32	0.15625
11/64	0.171875
3/16	0.1875
12/64	0.203125
7/32	0.21875
15/64	0.234375
1/4	0.25
17/64	0.265625
9/32	0.28125
19/64	0.296875
5/16	0.3125

Note: To find the decimal equivalent of a fraction, divide the numerator by the denominator.

AREA AND OTHER FORMULAS

Parallelogram	Area = base × distance between the two parallel sides
Pyramid	Area = 1/2 perimeter of base × slant height + area of base
	Volume = area of base × 1/3 of the altitude
Rectangle	Area = length × width
Rectangular prism	Volume = width × height × length
Sphere	Area of surface = diameter × diameter × 3.1416
	Side of inscribed cube = radius × 1.547
	Volume = diameter × diameter × diameter × 0.5236
Square	Area = length × width
Triangle	Area = one-half of height times base
Trapezoid	Area = one-half of the sum of the parallel sides × the height
Cone	Area of surface = one-half of circumference of base × slant height + area of base.
	Volume = diameter × diameter × 0.7854 × one-third of the altitude.
Cube	Volume = width × height × length
Ellipse	Area = short diameter × long diameter × 0.7854
Cylinder	Area of surface = diameter × 3.1416 × length + area of the two bases
	Area of base = diameter × diameter × 0.7854
	Area of base = volume ÷ length
	Length = volume ÷ area of base
	Volume = length × area of base
	Capacity in gallons = volume in inches ÷ 231
	Capacity of gallons = diameter × diameter × length × 0.0034
	Capacity in gallons = volume in feet × 7.48
Circle	Circumference = diameter × 3.1416
	Circumference = radius × 6.2832
	Diameter = radius × 2
	Diameter = square root of = (area ÷ 0.7854)
	Diameter = square root of area × 1.1283

AREA AND OTHER FORMULAS *(cont.)*

	Diameter = circumference × 0.31831
	Radius = diameter ÷ 2
	Radius = circumference × 0.15915
	Radius = square root of area × 0.56419
	Area = diameter × diameter × 0.7854
	Area = half of the circumference × half of the diameter
	Area = square of the circumference × 0.0796
	Arc length = degrees × radius × 0.01745
	Degrees of arc = length ÷ (radius × 0.01745)
	Radius of arc = length ÷ (degrees × 0.01745)
	Side of equal square = diameter × 0.8862
	Side of inscribed square = diameter × 0.7071
	Area of sector = area of circle × degrees of arc ÷ 360
Estimating volume	Multiply length × width × thickness
	Example: 50 ft × 10 ft × 8 in.
	$50 \times 10 \times 8/12 = 333.33^3$
	To convert to cubic yards, divide by 27. (There are 3 ft^3/yd^3)
	Example: $333.33 \div 27 = 12.35$ yd^3

PIPING

The capacity of pipes is as the square of their diameters. Thus, doubling the diameter of a pipe increases its capacity four times. The area of a pipe wall may be determined by the following formula:

Area of pipe wall = 0.7854 × [(o.d. × o.d.) − (i.d. × i.d.)]

The approximate weight of a piece of pipe may be determined by the following formulas:

Cast-iron pipe: weight = $(A^2 - B^2) \times$ length × 0.2042

Steel pipe: weight = $(A^2 - B^2) \times$ length × 0.2199

Copper pipe: weight = $(A^2 - B^2) \times$ length × 0.2537

A = outside diameter of the pipe in inches

B = inside diameter of the pipe in inches

TEMPERATURE CONVERSION

Temperature may be expressed according to the Fahrenheit (F) scale or the Celsius (C) scale. To convert °C to °F or °F to °C, use the following formulas:

$$°F = 1.8 \times °C + 32$$
$$°C = 0.55555555 \times °F - 32$$
$$°C = °F - 32 \div 1.8$$
$$°F = °C. \times 1.8 + 32$$

To figure the final temperature when two different temperatures of water are mixed together, use the following formula:

$$\frac{(A \times C) + (B \times D)}{A + B}$$

A = weight of lower temperature water
B = weight of higher temperature water
C = lower temperature
D = higher temperature

EXPANSION IN PLASTIC PIPING

The formula for calculating expansion or contraction in plastic piping is:

$$L = Y \times \frac{T - F}{10} \times \frac{L}{100}$$

L = expansion in inches
Y = constant factor expressing inches of expansion per 100 °F temperature change per 100 ft of pipe
T = maximum temperature (°F)
F = minimum temperature (°F)
L = length of pipe run in feet

RADIANT HEAT

RADIANT HEAT FACTS

Radiation

3 ft of 1-in. pipe equal 1 ft^2 R.
2⅓ lineal ft of 1¼-in. pipe equal 1 ft^2 R.
Hot water radiation gives off 150 Btu/ft^2 R/hr.
Steam radiation gives off 240 Btu/ft^2 R/hr.
On greenhouse heating, figure 2/3 ft^2 R/ft^2 glass.
1 ft^2 of direct radiation condenses 0.25 lb water/hr.

The formulas for pipe radiation of heat are as follows:

$$L = \frac{144}{OD \times 3.1416} \times R \div 12$$

D = outside diameter (OD) of pipe

L = length of pipe needed in feet

R = square feet of radiation needed

CALCULATING HEAT LOSS

The surface area of a material (in square feet) divided by its radiation (R) value multiplied by the difference in °Fahrenheit between inside and outside temperature equals heat loss in Btu/hr. Use the following formula:

$$(\text{Surface area} \div R \text{ value}) \times \begin{array}{l}(\text{temperature inside} - \\ \text{temperature outside})\end{array}$$

2

Reference Tables

Conversion tables make finding solutions to problems easy. If you need to convert a customary unit of measure to a metric measure, a conversion table is the fastest way to do it. Many times plumbers are required to convert various data to another form. You could spend hours learning all the knowledge needed to do this, or you can simply refer to the many conversion tables and formulas in this chapter. Look for yourself. You will find a table or formula for most any need that you come up against.

MEASUREMENT CONVERSION FACTORS

To Change	to	Multiply by
Inches	Feet	0.0833
Inches	Millimeters	25.4
Feet	Inches	12
Feet	Yards	0.3333
Yards	Feet	3
Square inches	Square feet	0.00694
Square feet	Square inches	144
Square feet	Square yards	0.11111
Square yards	Square feet	9
Cubic inches	Cubic feet	0.00058
Cubic feet	Cubic inches	1728
Cubic feet	Cubic yards	0.03703
Gallons	Cubic inches	231
Gallons	Cubic feet	0.1337
Gallons	Pounds of water	8.33
Pounds of water	Gallons	0.12004
Ounces	Pounds	0.0625
Pounds	Ounces	16
Inches of water	Pounds per square inch	0.0361
Inches of water	Inches of mercury	0.0735
Inches of water	Ounces per square inch	0.578
Inches of water	Pounds per square foot	5.2
Inches of mercury	Inches of water	13.6
Inches of mercury	Feet of water	1.1333
Inches of mercury	Pounds per square inch	0.4914
Ounces per square inch	Inches of mercury	0.127
Ounces per square inch	Inches of water	1.733
Pounds per square inch	Inches of water	27.72
Pounds per square inch	Feet of water	2.310
Pounds per square inch	Inches of mercury	2.04
Pounds per square inch	Atmospheres	0.0681
Feet of water	Pounds per square inch	0.434
Feet of water	Pounds per square foot	62.5
Feet of water	Inches of mercury	0.8824
Atmospheres	Pounds per square inch	14.696
Atmospheres	Inches of mercury	29.92
Atmospheres	Feet of water	34
Long tons	Pounds	2240
Short tons	Pounds	2000
Short tons	Long tons	0.89295

FORMULAS FOR CONCRETE

Grade	Ratio	Material Needed for Each Cubic Yard of Concrete
Strong—watertight, exposed to weather and moderate wear	1:2¼:3	6 bags cement 14 ft^3 sand (0.52 yd^3) 18 ft^3 stone (0.67 yd^3)
Moderate—moderate strength, not exposed	1:2¾:4	5 bags cement 14 ft^3 sand (0.52 yd^3) 20 ft^3 stone (0.74 yd^3)
Economy—massive areas, low strength	1:3:5	4½ bags cement 13 ft^3 sand (0.48 yd^3) 22 ft^3 stone (0.82 yd^3)

ROOF PITCHES

U.S.	Metric
2/12	50/300
4/12	100/300
6/12	150/300
8/12	200/300
10/12	250/300
12/12	300/300

MINUTES CONVERTED TO DECIMAL OF A DEGREE

Minutes	Decimal of a Degree	Minutes	Decimal of a Degree
1	0.0166	16	0.2666
2	0.0333	17	0.2833
3	0.0500	18	0.3000
4	0.0666	19	0.3166
5	0.0833	20	0.3333
6	0.1000	21	0.3500
7	0.1166	22	0.3666
8	0.1333	23	0.3833
9	0.1500	24	0.4000
10	0.1666	25	0.4166
11	0.1833		
12	0.2000		
13	0.2166		
14	0.2333		
15	0.2500		

METRIC CONVERSION TABLE

U.S.	Metric
0.001 in.	0.025 mm
1 in.	25.400 mm
1 ft	30.48 cm
1 ft	0.3048 m
1 yd	0.9144 m
1 mi	1.609 km
1 in.^{2}	6.4516 cm^2
1 ft^2	0.0929 m^2
1 yd^2	0.8361 m^2
1 a	0.4047 ha
1 mi^2	2.590 km^2
1 in.^{3}	16.387 cm^3
1 ft^3	0.0283 m^3
1 yd^3	0.7647 m^3
1 U.S. oz	29.57 ml
1 U.S. p	0.4732 l
1 U.S. gal	3.785 l
1 oz	28.35 g
1 lb	0.4536 kg

THERMAL EXPANSION OF PVC-DWV

	Temperature Change (°F)						
Length (ft)	40	50	60	70	80	90	100
20	0.278	0.348	0.418	0.487	0.557	0.626	0.696
40	0.557	0.696	0.835	0.974	1.114	1.235	1.392
60	0.835	1.044	1.253	1.462	1.670	1.879	2.088
80	1.134	1.392	1.670	1.879	2.227	2.506	2.784
100	1.192	1.740	2.088	2.436	2.784	3.132	3.480

THERMAL EXPANSION OF ALL PIPES (Except PVC-DWV)

	Temperature Change (°F)						
Length (ft)	40	50	60	70	80	90	100
20	0.536	0.670	0.804	0.938	1.072	1.206	1.340
40	1.070	1.340	1.610	1.880	2.050	2.420	2.690
60	1.609	2.010	2.410	2.820	3.220	3.620	4.020
80	2.143	2.680	3.220	3.760	4.290	4.830	5.360
100	2.680	3.350	4.020	4.700	5.360	6.030	6.700

CIRCUMFERENCE OF CIRCLE

Diameter	Circumference	Diameter	Circumference
1/8	0.3927	10	31.41
1/4	0.7854	10½	32.98
3/8	1.178	11	34.55
1/2	1.570	11½	36.12
5/8	1.963	12	37.69
3/4	2.356	12½	39.27
7/8	2.748	13	40.84
1	3.141	13½	42.41
1⅛	3.534	14	43.98
1¼	3.927	14½	45.55
1⅜	4.319	15	47.12
1½	4.712	15½	48.69
1⅝	5.105	16	50.26
1¾	5.497	16½	51.83
1⅞	5.890	17	53.40
2	6.283	17½	54.97
2¼	7.068	18	56.54
2½	7.854	18½	58.11
2¾	8.639	19	59.69
3	9.424	19½	61.26
3¼	10.21	20	62.83
3½	10.99	20½	64.40
3¾	11.78	21	65.97
4	12.56	21½	67.54
4½	14.13	22	69.11
5	15.70	22½	70.68
5½	17.27	23	72.25
6	18.84	23½	73.82
6½	20.42	24	75.39
7	21.99	24½	76.96
7½	23.56	25	78.54
8	25.13	26	81.68
8½	26.70	27	84.82
9	28.27	28	87.96
9 ½	29.84	29	91.10
		30	94.24

AREA OF CIRCLE

Diameter	Area	Diameter	Area
1/8	0.0123	10	78.54
1/4	0.0491	10½	86.59
3/8	0.1104	11	95.03
1/2	0.1963	11½	103.86
5/8	0.3068	12	113.09
3/4	0.4418	12½	122.71
7/8	0.6013	13	132.73
1	0.7854	13½	143.13
1⅛	0.9940	14	153.93
1¼	1.227	14½	165.13
1⅜	1.484	15	176.71
1½	1.767	15½	188.69
1⅝	2.073	16	201.06
1¾	2.405	16½	213.82
1⅞	2.761	17	226.98
2	3.141	17½	240.52
2¼	3.976	18	254.46
2½	4.908	18½	268.80
2¾	5.939	19	283.52
3	7.068	19½	298.60
3¼	8.295	20	314.16
3½	9.621	20½	330.06
3¾	11.044	21	346.36
4	12.566	21½	363.05
4½	15.904	22	380.13
5	19.635	22½	397.60
5½	23.758	23	415.47
6	28.274	23½	433.73
6½	33.183	24	452.39
7	38.484	24½	471.43
7½	44.178	25	490.87
8	50.265	26	530.93
8½	56.745	27	572.55
9	63.617	28	615.75
9½	70.882	29	660.52
		30	706.86

TEMPERATURE CONVERSION
−100°–30°

°C	Base Temperature	°F
−73	−100	−148
−68	−90	−130
−62	−80	−112
−57	−70	−94
−51	−60	−76
−46	−50	−58
−40	−40	−40
−34.4	−30	−22
−28.9	−20	−4
−23.3	−10	14
−17.8	0	32
−17.2	1	33.8
−16.7	2	35.6
−16.1	3	37.4
−15.6	4	39.2
−15.0	5	41.0
−14.4	6	42.8
−13.9	7	44.6
−13.3	8	46.4
−12.8	9	48.2
−12.2	10	50.0
−11.7	11	51.8
−11.1	12	53.6
−10.6	13	55.4
−10.0	14	57.2

31°–71°

°C	Base Temperature	°F
−0.6	31	87.8
0	32	89.6
0.6	33	91.4
1.1	34	93.2
1.7	35	95.0
2.2	36	96.8
2.8	37	98.6
3.3	38	100.4
3.9	39	102.2
4.4	40	104.0
5.0	41	105.8
5.6	42	107.6

TEMPERATURE CONVERSION (*cont.*)
31°–71°

°C	Base Temperature	°F
6.1	43	109.4
6.7	44	111.2
7.2	45	113.0
7.8	46	114.8
8.3	47	116.6
8.9	48	118.4
9.4	49	120.0
10.0	50	122.0
10.6	51	123.8
11.1	52	125.6
11.7	53	127.4
12.2	54	129.2
12.8	55	131.0

72°–212°

°C	Base Temperature	°F
22.2	72	161.6
22.8	73	163.4
23.3	74	165.2
23.9	75	167.0
24.4	76	168.8
25.0	77	170.6
25.6	78	172.4
26.1	79	174.2
26.7	80	176.0
27.8	81	177.8
28.3	82	179.6
28.9	83	181.4
29.4	84	183.2
30.0	85	185.0
30.6	86	186.8
31.1	87	188.6
31.7	88	190.4
32.2	89	192.2
32.8	90	194.0
33.3	91	195.8
33.9	92	197.6
34.4	93	199.4
35.0	94	201.2
35.6	95	203.0
36.1	96	204.8

TEMPERATURE CONVERSION *(cont.)*
213°–620°

°C	*Base Temperature*	°F
104	220	248
110	230	446
116	240	464
121	250	482
127	260	500
132	270	518
138	280	536
143	290	554
149	300	572
154	310	590
160	320	608
166	330	626
171	340	644
177	350	662
182	360	680
188	370	698
193	380	716
199	390	734
204	400	752
210	410	770
216	420	788
221	430	806
227	440	824
232	450	842
238	460	860

621°–1000°

°C	*Base Temperature*	°F
332	630	1166
338	640	1184
343	650	1202
349	660	1220
354	670	1238
360	680	1256
366	690	1274
371	700	1292
377	710	1310
382	720	1328

TEMPERATURE CONVERSION *(cont.)*
621°–1000°

°C	Base Temperature	°F
388	730	1346
393	740	1364
399	750	1382
404	760	1400
410	770	1418
416	780	1436
421	790	1454
427	800	1472
432	810	1490
438	820	1508
443	830	1526
449	840	1544
454	850	1562
460	860	1580
466	870	1598

VOLUME-TO-WEIGHT CONVERSIONS FOR SAND

1 ft^3	approx. 100 lb
1 yd^3	2700 lb
1 t	3/4 yd or 20 ft^3
Average shovelful	15 lb
12-qt pail	40 lb

USEFUL MULTIPLIERS

To Change	to	Multiply by
Inches	Feet	0.0833
Inches	Millimeters	25.4
Feet	Inches	12
Feet	Yards	0.3333
Yards	Feet	3
Square inches	Square feet	0.00694
Square feet	Square inches	144
Square feet	Square yards	0.11111
Square yards	Square feet	9
Cubic inches	Cubic feet	0.00058
Cubic feet	Cubic inches	1728
Cubic feet	Cubic yards	0.03703
Cubic yards	Cubic feet	27
Cubic inches	Gallons	0.00433
Cubic feet	Gallons	7.48
Gallons	Cubic inches	231
Gallons	Cubic feet	0.1337
Gallons	Pounds of water	8.33
Pounds of water	Gallons	0.12004
Ounces	Pounds	0.0625
Pounds	Ounces	16
Inches of water	Pounds per square inch	0.0361
Inches of water	Inches of mercury	0.0735
Inches of water	Ounces per square inch	0.578
Inches of water	Pounds per square foot	5.2
Inches of mercury	Inches of water	13.6
Inches of mercury	Feet of water	1.1333
Inches of mercury	Feet of water	0.4914
Ounces per square inch	Pounds per square inch	0.127
Ounces per square inch	Inches of mercury	1.733
Pounds per square inch	Inches of water	27.72
Pounds per square inch	Feet of water	2.310
Pounds per square inch	Inches of mercury	2.04
Pounds per square inch	Atmospheres	0.0681
Feet of water	Pounds per square inch	0.434
Feet of water	Pounds per square foot	62.5
Feet of water	Inches of mercury	0.8824
Atmospheres	Pounds per square inch	14.696
Atmospheres	Inches of mercury	29.92
Atmospheres	Feet of water	34
Long tons	Pounds	2240
Short tons	Pounds	2000
Short tons	Long tons	0.89285

SQUARE ROOTS OF NUMBERS

Number	Square Root	Number	Square Root	Number	Square Root
1	1.00000	36	6.00000	71	8.42614
2	1.41421	37	6.08276	72	8.48528
3	1.73205	38	6.16441	73	8.54400
4	2.00000	39	6.24499	74	8.60232
5	2.23606	40	6.32455	75	8.66025
6	2.44948	41	6.40312	76	8.71779
7	2.64575	42	6.48074	77	8.77496
8	2.82842	43	6.55743	78	8.83176
9	3.00000	44	6.63324	79	8.88819
10	3.16227	45	6.70820	80	8.94427
11	3.31662	46	6.78233	81	9.00000
12	3.46410	47	6.85565	82	9.05538
13	3.60555	48	6.92820	83	9.11043
14	3.74165	49	7.00000	84	9.16515
15	3.87298	50	7.07106	85	9.21954
16	4.00000	51	7.14142	86	9.27361
17	4.12310	52	7.21110	87	9.32737
18	4.24264	53	7.28010	88	9.38083
19	4.35889	54	7.34846	89	9.43398
20	4.47213	55	7.41619	90	9.48683
21	4.58257	56	7.48331	91	9.53939
22	4.69041	57	7.54983	92	9.59166
23	4.79583	58	7.61577	93	9.64365
24	4.89897	59	7.68114	94	9.69535
25	5.00000	60	7.74596	95	9.74679
26	5.09901	61	7.81024	96	9.79795
27	5.19615	62	7.87400	97	9.84885
28	5.29150	63	7.93725	98	9.89949
29	5.38516	64	8.00000	99	9.94987
30	5.47722	65	8.06225	100	10.00000
31	5.56776	66	8.12403		
32	5.65685	67	8.18535		
33	5.74456	68	8.24621		
34	5.83095	69	8.30662		
35	5.91607	70	8.36660		

CUBES OF NUMBERS

Number	Cube	Number	Cube	Number	Cube
1	1	36	46,656	71	357,911
2	8	37	50,653	72	373,248
3	27	38	54,872	73	389,017
4	64	39	59,319	74	405,224
5	125	40	64,000	75	421,875
6	216	41	68,921	76	438,976
7	343	42	74,088	77	456,533
8	512	43	79,507	78	474,552
9	729	44	85,184	79	493,039
10	1,000	45	91,125	80	512,000
11	1,331	46	97,336	81	531,441
12	1,728	47	103,823	82	551,368
13	2,197	48	110,592	83	571,787
14	2,477	49	117,649	84	592,704
15	3,375	50	125,000	85	614,125
16	4,096	51	132,651	86	636,056
17	4,913	52	140,608	87	658,503
18	5,832	53	148,877	88	681,472
19	6,859	54	157,464	89	704,969
20	8,000	55	166,375	90	729,000
21	9,621	56	175,616	91	753,571
22	10,648	57	185,193	92	778,688
23	12,167	58	195,112	93	804,357
24	13,824	59	205,379	94	830,584
25	15,625	60	216,000	95	857,375
26	17,576	61	226,981	96	884,736
27	19,683	62	238,328	97	912,673
28	21,952	63	250,047	98	941,192
29	24,389	64	262,144	99	970,299
30	27,000	65	274,625	100	1,000,000
31	29,791	66	287,496		
32	32,768	67	300,763		
33	35,937	68	314,432		
34	39,304	69	328,500		
35	42,875	70	343,000		

SQUARES OF NUMBERS

Number	Square	Number	Square	Number	Square
1	1	36	1296	71	5041
2	4	37	1369	72	5184
3	9	38	1444	73	5329
4	16	39	1521	74	5476
5	25	40	1600	75	5625
6	36	41	1681	76	5776
7	49	42	1764	77	5929
8	64	43	1849	78	6084
9	81	44	1936	79	6241
10	100	45	2025	80	6400
11	121	46	2116	81	6561
12	144	47	2209	82	6724
13	169	48	2304	83	6889
14	196	49	2401	84	7056
15	225	50	2500	85	7225
16	256	51	2601	86	7396
17	289	52	2704	87	7569
18	324	53	2809	88	7744
19	361	54	2916	89	7921
20	400	55	3025	90	8100
21	441	56	3136	91	8281
22	484	57	3249	92	8464
23	529	58	3364	93	8649
24	576	59	3481	94	8836
25	625	60	3600	95	9025
26	676	61	3721	96	8216
27	729	62	3844	97	9409
28	784	63	3969	98	9604
29	841	64	4096	99	9801
30	900	65	4225	100	10000
31	961	66	4356		
32	1024	67	4489		
33	1089	68	4624		
34	1156	69	4761		
35	1225	70	4900		

POUNDS TO KILOGRAMS

Pounds	Kilograms
1	0.454
2	0.907
3	1.361
4	1.814
5	2.268
6	2.722
7	3.175
8	3.629
9	4.082
10	4.536
25	11.34
50	22.68
75	34.02
100	45.36

OUNCES TO KILOGRAMS

Ounces	Kilograms
1	0.028
2	0.057
3	0.085
4	0.113
5	0.142
6	0.170
7	0.198
8	0.227
9	0.255
10	0.283
11	0.312
12	0.340
13	0.369
14	0.397
15	0.425
16	0.454

FLOW-RATE CONVERSION

Gallons per Minute	Liters per Minute
1	3.75
2	6.50
3	11.25
4	15.00
5	18.75
6	22.50
7	26.25
8	30.00
9	33.75
10	37.50

FLOW-RATE EQUIVALENTS

1 GMP   0.134 cu ft/min

1 cu ft/min (cfm)   448.8 gal/hr (gph)

Feet per Second	Meters per Second
1	0.3050
2	0.610
3	0.915
4	1.220
5	1.525
6	1.830
7	2.135
8	2.440
9	2.754
10	3.050

POUNDS PER SQUARE FOOT TO KILOPASCALS

Pounds per Square Foot	Kilopascals
1	0.0479
2	0.0958
3	0.1437
4	0.1916
5	0.2395
6	0.2874
7	0.3353
8	0.3832
9	0.4311
10	0.4788
25	1.1971
50	2.394
75	3.5911
100	4.7880

POUNDS PER SQUARE INCH TO KILOPASCALS

Pounds per Square Inch	Kilopascals
1	6.895
2	13.790
3	20.685
4	27.580
5	34.475
6	41.370
7	48.265
8	55.160
9	62.055
10	68.950
25	172.375
50	344.750
75	517.125
100	689.500

DECIMAL EQUIVALENTS OF AN INCH

1/32	0.03125
1/16	0.0625
3/32	0.09375
1/8	0.125
5/32	0.15625
3/16	0.1875
7/32	0.21875
1/4	0.25
9/32	0.28125
5/16	0.3125
11/32	0.34375
3/8	0.375
13/32	0.40625
7/16	0.4375
15/32	0.46875
1/2	0.500
17/32	0.53125
9/16	0.5625
19/32	0.59375
5/8	0.625
21/32	0.65625
11/16	0.6875
23/32	0.71875
3/4	0.75
25/32	0.78125
13/16	0.8125
27/32	0.84375
7/8	0.875
29/32	0.90625
15/16	0.9375
31/32	0.96875
1	1.000

DECIMALS TO MILLIMETERS

Decimal Equivalent	Millimeters
0.0625	1.59
0.1250	3.18
0.1875	4.76
0.2500	6.35
0.3125	7.94
0.3750	9.52
0.4375	11.11
0.5000	12.70
0.5625	14.29
0.6250	15.87
0.6875	17.46
0.7500	19.05
0.8125	20.64
0.8750	22.22
0.9375	23.81
1.000	25.40

FRACTIONS TO DECIMALS

Fractions	Decimal Equivalent
1/16	0.0625
1/8	0.1250
3/16	0.1875
1/4	0.2500
5/16	0.3125
3/8	0.3750
7/16	0.4375
1/2	0.5000
9/16	0.5625
5/8	0.6250
11/16	0.6875
3/4	0.7500
13/16	0.8125
7/8	0.8750
15/16	0.9375
1	1.000

DECIMAL EQUIVALENTS OF FRACTIONS

Fraction	Decimal
1/64	0.015625
1/32	0.03125
3/64	0.046875
1/20	0.05
1/16	0.0625
1/13	0.0769
5/64	0.078125
1/12	0.0833
1/11	0.0909
3/32	0.09375
1/10	0.10
7/64	0.109375
1/9	0.111
1/8	0.125
9/64	0.140625
1/7	0.1429
5/32	0.15625
1/6	0.1667
11/64	0.171875
3/16	0.1875
1/5	0.2
13/64	0.203125
7/32	0.21875
15/64	0.234375
1/4	0.25
17/64	0.265625
9/32	0.28125
19/64	0.296875
5/16	0.3125
21/64	0.328125
1/3	0.333
11/32	0.34375
23/64	0.359375
3/8	0.375
25/64	0.390625
13/32	0.40625
27/64	0.421875
7/16	0.4375
29/64	0.453125
15/32	0.46875
31/64	0.484375
1/2	0.5
33/64	0.515625

DECIMAL EQUIVALENTS OF FRACTIONS *(cont.)*

Fraction	Decimal
17/32	0.53125
35/64	0.546875
9/16	0.5625
37/64	0.578125
19/32	0.59375
39/64	0.609375
5/8	0.625
41/64	0.640625
21/32	0.65625
43/64	0.671875
11/16	0.6875
45/64	0.703125

WATER FEET HEAD TO POUNDS PER SQUARE INCH

Feet Head	Pounds per Square Inch	Feet Head	Pounds per Square Inch
1	0.43	50	21.65
2	0.87	60	25.99
3	1.30	70	30.32
4	1.73	80	34.65
5	2.17	90	38.98
6	2.60	100	43.34
7	3.03	110	47.64
8	3.46	120	51.97
9	3.90	130	56.30
10	4.33	140	60.63
15	6.50	150	64.96
20	8.66	160	69.29
25	10.83	170	73.63
30	12.99	180	77.96
40	17.32	200	86.62

WATER PRESSURE IN POUNDS
WITH EQUIVALENT FEET HEADS

Pounds per Square Inch	Feet Head
1	2.31
2	4.62
3	6.93
4	9.24
5	11.54
6	13.85
7	16.16
8	18.47
9	20.78
10	23.09
15	34.63
20	46.18
25	57.72
30	69.27
40	92.36
50	115.45
60	138.54
70	161.63
80	184.72
90	207.81
100	230.90
110	253.98
120	277.07
130	300.16
140	323.25
150	346.34
160	369.43
170	392.52
180	415.61
200	461.78
250	577.24
300	692.69
350	808.13
400	922.58
500	1154.48
600	1385.39
700	1616.30
800	1847.20
900	2078.10
1000	2309.00

DESIGN TEMPERATURE

Outside design temperature = average of lowest recorded temperature in each month from October to March

Inside design temperature = 70°F or as specified by owner

A degree day is one day multiplied by the number of Fahrenheit degrees the mean temperature is below 65°F. The number of degree days in a year is a good guideline for designing heating and insulation systems.

FACTORS USED IN CONVERTING FROM CUSTOMARY (U.S.) UNITS TO METRIC UNITS

To Find	Multiply	By
Microns	Mils	25.4
Centimeters	Inches	2.54
Meters	Feet	0.3048
Meters	Yards	0.19144
Kilometers	Miles	1.609344
Grams	Ounces	28.349523
Kilograms	Pounds	0.4539237
Liters	Gallons (U.S.)	3.7854118
Liters	Gallons (imperial)	4.546090
Milliliters (cc)	Fluid ounces	29.573530
Milliliters (cc)	Cubic inches	16.387064
Square centimeters	Square inches	6.4516
Square meters	Square feet	0.09290304
Square meters	Square yards	0.83612736
Cubic meters	Cubic feet	2.8316847×10^{-2}
Cubic meters	Cubic yards	0.76455486
Joules	Btu	1054.3504
Joules	Foot-pounds	1.35582
Kilowatts	Btu per minute	0.01757251
Kilowatts	Foot-pounds per minute	2.2597×10^{-5}
Kilowatts	Horsepower	0.7457
Radians	Degrees	0.017453293
Watts	Btu per minute	17,5725

METRIC MOTOR RATINGS

Horsepower	Kilowatt
1/20	0.025
1/16	0.05
1/8	0.1
1/6	0.14
1/4	0.2
1/3	0.28
1/2	0.4
1	0.8
1½	1.1

MOTOR POWER OUTPUT COMPARISON

Watt Output	Millihorsepower (MHP)	Fractional HP
0.746	1	1/1000
1.492	2	1/500
2.94	4	1/250
4.48	6	1/170
5.97	8	1/125
7.46	10	1/100
9.33	12.5	1/80
10.68	14.3	1/70
11.19	15	1/65
11.94	16	1/60
14.92	20	1/50
18.65	25	1/40
22.38	30	1/35
24.90	33	1/30

SQUARE MEASURE

Metric	U.S.
144 in.^{2}	1 ft^2
9 ft^2	1 yd^2
1 yd^2	1296 in.^{2}
4840 yd^2	1 a
640 a	1 mi^2

SQUARE MEASURES OF LENGTH AND AREA

$1\ cm^2$	$0.1550\ in.^2$
$1\ dm^2$	$0.1076\ ft^2$
$1\ m^2$	$1.196\ yd^2$
$1\ A\,(are)$	$3.954\ rd^2$
$1\ ha$	$2.47\ a\,(acres)$
$1\ km^2$	$0.386\ mi^2$
$1\ in.^2$	$6.452\ cm^2$
$1\ ft^2$	$9.2903\ dm^2$
$1\ yd^2$	$0.8361\ m^2$
$1\ rd^2$	$0.2529\ A\,(are)$
$1\ a\,(acre)$	$0.4047\ ha$
$1\ mi^2$	$2.59\ km^2$

WATER WEIGHT

$1\ ft^3$ at 50°F weighs 62.41 lb
1 gal at 50°F weighs 8.34 lb
$1\ ft^3$ of ice weighs 57.2 lb
Water is at its greatest density at 39.2°F
$1\ ft^3$ at 39.2°F weighs 62.43 lb

WATER VOLUME TO WEIGHT

$1\ ft^3$	62.4 lbs
$1\ ft^3$	7.48 gal
1 gal	8.33 lbs
1 gal	$0.1337\ ft^3$

METRIC ABBREVIATIONS

Unit	Symbol
Length	
Millimeter	mm
Centimeter	cm
Meter	m
Kilometer	km
Area	
Square millimeter	mm^2
Square centimeter	cm^2
Square decimeter	dm^2
Square meter	m^2
Square kilometer	km^2
Volume	
Cubic centimeter	cm^3
Cubic decimeter	dm^3
Cubic meter	m^3
Mass	
Milligram	mg
Gram	g
Kilogram	kg
Tonne	t
Temperature	
Degrees Celsius	°C
Kelvin	K
Time	
Second	s
Plane angle	
Radius	rad
Force	
Newton	N
Energy, work, quantity of heat	
Joule	J
Kilojoule	kJ
Megajoule	MJ
Power, heat flow rate	
Watt	W
Kilowatt	kW
Pressure	
Pascal	Pa
Kilopascal	kPa
Megapascal	MPa
Velocity, speed	
Meter per second	m/s
Kilometer per hour	km/h

MEASUREMENT CONVERSIONS (IMPERIAL TO METRIC)

Length	
1 in.	25.4 mm
1 ft	0.3048 m
1 yd	0.9144 m
1 mi	1.609 km
Mass	
1 lb	0.454 kg
1 U.S. short ton	0.9072 t
Area	
1 ft^2	0.092 m^2
1 yd^2	0.836 m^2
1 a	0.404 ha
Capacity, liquid	
1 ft^3	0.028 m^3
Capacity, dry	
1 yd^3	0.764 m^3
Volume, liquid	
1 qt	0.946 l
1 gal	3.785 l
Heat	
1 Btu	1055 joule (J)
1 Btu/hr	0.293 watt (W)

AVOIRDUPOIS WEIGHT

16 oz = 1 lb
100 lb = 1 cwt
20 cwt = 1 ton
1 ton = 2000 lb
1 long ton = 2240 lb

16 dr = 1 oz
16 oz = 1 lb
100 lb = 1 cwt
20 cwt = 1 ton = 2000 lb

METRIC WEIGHT MEASURE

10 milligrams (mg)	1 centigram
10 centigrams (cg)	1 decigram
10 decigrams (dg)	1 gram
10 grams (g)	1 dekagram
10 dekagrams (dkg)	1 hectogram
10 hectograms (hg)	1 kilogram
10 kilogram (kg)	1 myriagram
10 myriagrams (myg)	1 quintal

VOLUME MEASURE EQUIVALENTS

1 gal	0.133681 ft^3
1 gal	231 in.^3

CIRCULAR MEASURES

60 s	1 min
60 min	1°
360°	1 circle

MEASURES OF LENGTH

12 in.	1 ft	
3 ft	1 yd	
5½ yd or 16½ ft	1 rd	
40 rd	1 fur	
8 fur or 320 rd	1 mi	
	(sometimes called statute mile)	
12 in	1 ft	
3 ft	1 yd	36 in.
5½ yd	1 rd	16½ ft
40 rd	1 fur	660 ft
8 fur	1 mi	5280 ft

LIQUID MEASUREMENT EQUIVALENTS

4 gi	1 pt	28.875 in.^{3}
2 pt	1 qt	57.75 in.^{3}
4 qt	1 gal	231 in.^{3}
31½ gal	1 bbl	—
1 gal	231 in^3	—
7.48 gal	1 ft^3	—
1 gal water	8.33 lb	—
1 gal gasoline	5.84 lb	—

SURFACE MEASURES

Metric	U.S.
1 m^2	10.764 ft^2 (1.196 yd^2)
1 cm^2	0.155 in.^{2}
1 mm^2	0.00155 in.^{2}
0.836 m^2	1 yd^2
0.0929 m^2	1 ft^2
6.452 cm^2	1 in.^{2}
645.2 mm^2	1 in.^{2}

SQUARE MEASURES

144 in.^{2}	1 ft^2	
9 ft^2	1 yd^2	
30¼ yd^2	1 rd^2	272.25 ft^2
160 rd^2	1 a	4840 yd^2 (43,560 ft^2)
640 a	1 m^2	3,097,600 yd^2
36 mi^2	1 township	

SURVEYOR'S MEASURE

7.92 in.	1 link	
100 links	1 chain	66 ft
10 chains	1 fur	660 ft
80 chains	1 mi	5280 ft

SURFACE MEASURE

144 in.^{2}	1 ft^2
9 ft^2	1 yd^2
30¼ yd^2	1 rd^2
160 rd^2	1 a
640 a	1 mi^2
43,560 ft^2	1 a

CUBIC VOLUME

1728 in.^{3}	1 ft^3
27 ft^3	1 yd^3

DRY MEASURES

2 pt	1 qt	67.2 in.^{3}
8 qt	1 pk	537.61 in.^{3}
4 pk	1 bu	2150.42 in.^{3}

METRIC MEASURE

Cubic Measure

1000 cubic millimeters (mm^3)	1 cubic centimeter (cm^3)
1000 cubic centimeters (cm^3)	1 cubic decimeter (dm^3)
1000 cubic decimeters (dm^3)	1 cubic meter (m^3)

Capacity Measure

10 milliliters (ml)	1 centiliter (cl)
10 centiliters (cl)	1 deciliter (dl)
10 deciliters (dl)	1 liter (l)
10 liters (l)	1 dekaliter (dkl)
10 dekaliters (dkl)	1 hectoliter (hl)
10 hectoliters (hl)	1 kiloliter (kl)
10 kiloliters (kl)	1 myrialiter (myl)

WEIGHT EQUIVALENTS

1 g	0.0527 oz
1 kg	2.2046 lb
1 t	1.1023 long t
1 oz	28.35 g
1 lb	0.4536 kg
1 long t	0.9072 t

AREA IN INCHES AND MILLIMETERS

Square Inches	Square Millimeters
0.01227	8.0
0.04909	31.7
0.11045	71.3
0.19635	126.7
0.44179	285.0
0.7854	506.7
1.2272	791.7
1.7671	1140.1
3.1416	2026.8
4.9087	3166.9
7.0686	4560.4
12.566	8107.1
19.635	12,667.7
28.274	18,241.3
38.485	24,828.9
50.265	32,478.9
63.617	41,043.1
78.540	50,670.9

CIRCUMFERENCE IN INCHES AND MILLIMETERS

Inches	Millimeters
0.3927	10
0.7854	20
1.1781	30
1.5708	40
2.3562	60
3.1416	80
3.9270	100
4.7124	120
6.2832	160
7.8540	200
9.4248	240
12.566	320
15.708	400
18.850	480
21.991	560
25.133	640
28.274	720
31.416	800

DIAMETER IN INCHES AND MILLIMETERS

Inches	Millimeters
1/8	3.2
1/4	6.4
3/8	9.5
1/2	12.7
3/4	19.1
1	25.4
1¼	31.8
1½	38.1
2	50.8
2½	63.5
3	76.2
4	101.6
5	127
6	152.4
7	177.8
8	203.2
9	228.6
10	254

FEET TO METERS AND MILLIMETERS

Feet	Meters	Millimeters
1	0.305	304.8
2	0.610	609.6
3 (1 yd)	0.914	914.4
4	1.219	1 219.2
5	1.524	1 524.0
6 (2 yd)	1.829	1 828.8
7	2.134	2 133.6
8	2.438	2 438.2
9 (3 yd)	2.743	2 743.2
10	3.048	3 048.0
20	6.096	6 096.0
30 (10 yd)	9.144	9 144.0
40	12.19	12 192.0
50	15.24	15 240.0
60 (20 yd)	18.29	18 288.0
70	21.34	21 336.0
80	24.38	24 384.0
90 (30 yd)	27.43	27 432.0
100	30.48	30 480.0

ENGLISH TO METRIC CONVERSIONS

inches × 25.4	millimeters
feet × 0.3048	meters
miles × 1.6093	kilometers
square inches × 6.4515	square centimeters
square feet × 0.09290	square meters
acres × 0.4047	hectares
acres × 0.00405	square kilometers
cubic inches × 16.3872	cubic centimeters
cubic feet × 0.02832	cubic meters
cubic yards × 0.76452	cubic meters
cubic inches × 0.01639	liters
U.S. gallons × 3.7854	liters
ounces (avoirdupois) × 28.35	grams
pounds × 0.4536	kilograms
pounds per square inch (psi) × 0.0703	kilograms per square centimeter
pounds per cubic foot × 16.0189	kilograms per cubic meter
tons (2000 lb) × 0.9072	metric tons (1000 kg)
horsepower × 0.746	kilowatts

APPROXIMATE METRIC EQUIVALENTS

1 dmr	4 in.
1 m	1.1 yd
1 km	5/8 mi
1 ha	2½ a
1 stere(s) (or m^3)	1/4 cd
1 l	1.06 qt liquid; 0.9 qt dry
1 hl	2⅝ bu
1 kg	2⅕ lb
1 t	2200 lb

LENGTHS IN METRIC EQUIVALENTS

1 mm	1 mm	1/1000 m	0.03937 in
10 mm	1 cm	1/100 m	0.3937 in.
10 cm	1 dm	1/10 m	3.937 in.
10 dm	1 m	1 m	39.37 in.
10 m	1 dkm	10 m	32.8 ft
10 dkm	1 hm	100 m	328.09 ft
10 hm	1 km	1000 m	0.62137 mi

INCHES TO MILLIMETERS

Inches	Millimeters
1	25.4
2	50.8
3	76.2
4	101.6
5	127.0
6	152.4
7	177.8
8	203.2
9	228.6
10	254.0
11	279.4
12	304.8
13	330.2
14	355.6
15	381.0
16	406.4
17	431.8
18	457.2
19	482.6
20	508.0

SQUARE FEET TO SQUARE METERS

Square Feet	Square Meters
1	0.925
2	0.1850
3	0.2775
4	0.3700
5	0.4650
6	0.5550
7	0.6475
8	0.7400
9	0.8325
10	0.9250
25	2.315
50	4.65
100	9.25

METRIC WEIGHT CONVERSIONS

1 g	15.432 gr
1 kg	2.2046 lb
1 t	0.9842 t of 2240 lb
	19.68 cwt
	2204.6 lb
1 gr	0.0648 g
1 oz avdp	28.35 g
1 lb	0.4536 kg
1 t of 2240 lb	1.1016 t
	1016 kg

METRIC SQUARE MEASURE

100 mm^2	1 cm^2
100 cm^2	1 dm^2
100 dm^2	1 m^2

METRIC LINEAR MEASURE

	1 mm	0.001 m
10 mm	1 cm	0.01 m
10 cm	1 dm	0.1 m
10 dm	1 m	1 m
10 m	1 dkm	10 m
10 dkm	1 hm	100 m
10 hm	1 km	1000 m
10 km	1 mym	10,000 m

METRIC LAND MEASURE

1 ca	1 m^2
100 ca	1 A (are)
100 A (ares)	1 ha
100 ha	1 km^2

METRIC EQUIVALENTS

1 m	39.3 in.
	3.28083 ft
	1.0936 yd
1 cm	0.3937 in.
1 mm	0.03937 in., or nearly 1/25 in.
1 km	0.62137 mi
0.3048 m	1 ft
2.54 cm	1 in.
	25.40 mm

INCHES TO METERS AND MILLIMETERS

Inches	*Meters*	*Millimeters*
1/8	0.003	3.17
1/4	0.006	6.35
3/8	0.010	9.52
1/2	0.013	12.6
5/8	0.016	15.87
3/4	0.019	19.05
7/8	0.022	22.22
1	0.025	25.39
2	0.051	50.79
3	0.076	76.20
4	0.102	101.6
5	0.127	126.9
6	0.152	152.4
7	0.178	177.8
8	0.203	203.1
9	0.229	228.6
10	0.254	253.9
11	0.279	279.3
12	0.305	304.8

3

Sizing Pipe for Sanitary Drains

Sizing pipe is a task that most residential plumbers are required to tackle on a frequent basis. Commercial jobs are often sized by engineers and architects, but this is not the case with residential work. It is normally the plumber who determines what size pipes to install when plumbing a house. If you make a mistake, correcting it can be costly, not to mention a tad embarrassing.

Your local code book contains sizing tables and charts. Many times these tools speak for themselves, but sometimes they can be confusing, especially if you are not accustomed to working with them. Although it is true that your sizing must conform to code regulations, some rule-of-thumb methods often work very well. The combined use of experience, charts, and tables normally results in a satisfactory design. You're going to get a little of each in this chapter, in which I provide you with sizing tables, as well as some methods that experience has taught me.

SIZING BUILDING DRAINS AND SEWERS

The sizes of building drains and sewers are determined in the same way that proper pipe sizes are determined. The two components you must know to enable you to size these types of pipes are the total number of drainage fixture-units entering the pipe and the amount of fall placed on the pipe. The amount of fall is based on how much the pipe drops in each foot it travels. A normal grade is generally 1/4 in. to the foot, but the fall could be more or less.

When you refer to your code book, you will find information, probably a table, to aid you is sizing building drains and sewers. Let's take a look at how a building drain for a typical house would be sized in zone three.

Our sample house has 2½ bathrooms, a kitchen, and a laundry room. To size the building drain for this house, we must determine the total fixture-unit load that may be placed on the building drain. To do this, we start by listing all the plumbing fixtures producing a drainage load. In this house we have the following fixtures:

One bathtub	One shower
Three toilets	Three lavatories
One kitchen sink	One dishwasher
One clothes washer	One laundry tub

By using a chart in the code book, we can determine the number of drainage fixture-units assigned to each of these fixtures. When we add all the fixture-units, we have a total load of 28 fixture units. It is always best to allow a little extra in your fixture-unit load, so that your pipe will be in no danger of becoming overloaded. The next step is to look in your code book to determine the minimum drainage-pipe pitch allowed.

Our building drain will be installed with a 1/4-in. fall. By looking at the tables, we see that we can use a 3-in. pipe for our building drain, based on the number of fixture-units, but we notice that a 3-in. pipe may not carry the discharge of more than two toilets, and our test house has three toilets. This means we will have to move up to a 4-in. pipe.

Suppose our test house had only two toilets. What would the outcome be then? If we eliminate one of the toilets, our fixture-unit load drops to 24. According to the table, we could use a 2½-in. pipe, but we know our building drain must be at least a 3-in. pipe to connect to the toilets. A fixture's drain may enter a pipe that is the same size as the fixture drain or a pipe that is larger, but it may never be reduced to a smaller size, except with a 4-by-3-in. closet-bend.

HORIZONTAL BRANCHES

Horizontal branches are the pipes branching off from a stack that accept the discharge from fixture drains. These horizontal branches normally leave the stack as a horizontal pipe, but they may turn to a vertical position while retaining the name of horizontal branch.

The procedure for sizing a horizontal branch is similar to that used to size a building drain or sewer, but the ratings are different. Your code book will contain the benchmarks for your sizing efforts, but let me give you some examples.

The number of fixture-units allowed on a horizontal branch is determined by the size and pitch of the pipes. All of the following examples are based on a pitch of 1/4-in. to the foot. A 2-in. pipe can accommodate up to six fixture-units except in zone one, where it can have eight fixture-units. A 3-in. pipe can handle 20 fixture-units but not more than two toilets. In zone one, a 3-in. pipe is allowed up to 35 units and up to three toilets. A pipe of 1½ in. will carry three fixture-units unless you are in zone one. Zone one allows only a pipe of 1½ in. to carry two fixture-units, and they may not be from sinks, dishwashers, or urinals. A 4-in. pipe will take up to 160 fixture-units, except in zone one, where it will take up to 216 units.

Tables are provided in plumbing codes to make pipe sizing uniform and easier to do. Look over the tables provided here to see the types of information used to determine the size requirements for drainage systems.

SIZING STACKS

	Pipe Size (inches)	Number of Fixture-Units That Discharge on Stack from a Branch	Total Fixture-Units Allowed on Stack
Zone 2	1½	3	4
	2	6	10
	3	20[a]	30[a]
	4	160	240
Zone 3	1½	2	4
	2	6	10
	3	20[a]	48[a]
	4	90	240

[a]No more than two toilets may be placed on a 3-in. branch, and no more than six toilets may be connected to a 3-in. stack.

SIZING TALL STACKS WITH MORE THAN THREE BRANCH INTERVALS

	Pipe Size (inches)	Number of Fixture-Units That Discharge on Stack from a Branch	Total Fixture-Units Allowed on Stack
Zone 2	1½	2	8
	2	6	24
	3	16[a]	60[a]
	4	90	500
Zone 3	1½	2	8
	2	6	24
	3	20[a]	72[a]
	4	90	500

[a]No more than two toilets may be placed on a 3-in. branch, and no more than six toilets may be connected to a 3-in. stack.

EXAMPLE OF SIZING A HORIZONTAL BRANCH IN ZONE TWO

Pipe Size (inches)	Maximum Number of Fixture-Units on a Horizontal Branch
1¼	1
1½	3
2	6
3	20[a]
4	160
6	620

[a]Not more than two toilets may be connected to a single 3-in. horizontal branch. Any branch connecting with a toilet must have a minimum diameter of 3 in.

[b]Table does not represent branches of the building drain, and other restrictions apply under battery-venting conditions.

SIZING BUILDING DRAINS IN ZONE THREE

Pipe Size (inches)	Pipe Grade to the Foot (inches)	Maximum Number of Fixture-Units
2	1/4	21
3	1/4	42[a]
4	1/4	216

[a]No more than two toilets may be installed on a 3-in. building drain.

FIXTURE-UNIT REQUIREMENTS FOR TRAP SIZES

	Trap Size (inches)	Number of Fixture-Units
Zone 1	1¼	1
	1½	3
	2	4
	3	6
	4	8
Zone 2	1¼	1
	1½	2
	2	3
	3	5
	4	6
Zone 3	1¼	1
	1½	2
	2	3
	3	5
	4	6

RECOMMENDED TRAP SIZES

Type of Fixture	Zone 1 Trap Size (inches)	Zone 2 Trap Size (inches)	Zone 3 Trap Size (inches)
Bathtub	1½	1½	1½
Shower	2	2	2
Residential toilet	Integral	Integral	Integral
Lavatory	1¼	1¼	1¼
Bidet	1½	1½	1¼
Laundry tub	1½	1½	1½
Washing-machine standpipe	2	2	2
Floor drain	2	2	2
Kitchen sink	1½	1½	1½
Dishwasher	1½	1½	1½
Drinking fountain	1¼	1	1¼
Public toilet	Integral	Integral	Integral
Urinal			2

4

Sizing Vent Pipes

Sizing vent pipes is similar to sizing drainage systems. To size vent systems, you must understand the various types of vents. This chapter introduces you to the most common types of vents and the ways to design pipe sizes to create a vent system. The text, when combined with the tables provided, makes it easier to understand the methods used to arrive at pipe sizes for vent systems.

INDIVIDUAL VENTS

Individual vents are, as the name implies, vents that serve individual fixtures. These vents vent only one fixture, but they may connect into another vent that extends to the open air. Individual vents do not have to extend directly from the fixture being served to the outside air, without joining another part of the venting system, but eventually they must vent to open air space.

Sizing an individual vent is easy. The vent must be at least half the size of the drain it serves, but it may not have a diameter of less than 1¼ in. For example, a vent for a 3-in. drain could, in most cases, have a 1½-in. diameter. A vent for a 1½-in. drain may not have a diameter of less than 1¼ in.

RELIEF VENTS

Relief vents are used in conjunction with other vents. Their purpose is to provide additional air to the drainage system when the primary vent is too far from the fixture. A relief vent must be at least half the size of the pipe it is venting. For example, if a relief vent vents a 3-in. pipe, the relief vent must have a diameter of 1½ in., or larger.

CIRCUIT VENTS

Circuit vents are used with a battery of plumbing fixtures. Circuit vents are normally installed just before the last fixture of the battery. Then the circuit vent is extended upward to the open air or tied into another vent that extends to the outside. Circuit vents may tie into stack vents or vent stacks. When sizing a circuit vent, you must account for its developed length. In any event, the diameter of a circuit vent must be at least half the size of the drain it is serving.

DEVELOPED LENGTH

What effect does the length of the vent have on the vent's size? The developed length, the total linear footage of pipe making up the vent, is used in conjunction with factors provided in code books to determine vent sizes. To size circuit vents, branch vents, and individual vents for horizontal drains, you must use this method of sizing.

The criteria needed for sizing a vent, based on developed length, are the grade of the drainage pipe, the size of the drainage pipe, the developed length of the vent, and the factors allowed by local code requirements. Let's look at a few examples of how to size a vent by using this method.

For our first example, assume that the drain you are venting is a 3-in. pipe with a grade of 1/4 in./ft. This sizing exercise is done using zone three requirements. After looking at the tables and charts in the code book, we find that a 3-in. drain, running horizontally, with a grade of 1/4 in./ft, can be vented with a 1½-in. vent that has a developed length of 97 ft. It would be rare to extend a vent anywhere near 97 ft, but if your vent needed to exceed this distance, you could go to a larger vent. A 2-in. vent would allow you to extend the vent for a total length of 420 ft. A vent larger than 2 in. would allow you to extend the vent indefinitely.

For the second example, still using zone three's rules, assume that the drain is a 4-in. pipe, with a grade of 1/4 in./ft. In this case you could not use a 1½-in. vent. Remember, the vent must be at least half the size of the drain it is venting. A 2-in. vent would allow a developed vent length of 98 ft, and a 3-in. vent would allow the vent to extend to an unlimited length. As you can see, this type of sizing is not difficult.

Now, let's size a vent by using zone one's rules. In zone one, vent sizing is based on the vent's length and the number of fixture-units on the vent. If you were sizing a vent for a lavatory, you would need to know how many fixture-units the lavatory represents. Lavatories are rated as one fixture-unit. By using a table in the code book, you would find that a vent serving one fixture-unit can have a diameter of 1¼ in. and extend for 45 ft. A bathtub, rated at two fixture-units, would require a 1½-in. vent. The bathtub vent could run for 60 ft.

BRANCH VENTS

Branch vents extend horizontally and connect multiple vents. Branch vents are sized by using the developed-length method, just as you were shown in the preceding examples. A branch vent or an individual vent that is the same size as the drain it serves is unlimited in the developed length it may obtain. Be advised, however, that zone two and zone three use different tables and ratings for sizing various types of vents, whereas zone one uses the same rating and table for all normal venting situations.

VENT STACKS

A vent stack is a pipe used only for the purpose of venting. Vent stacks extend upward from the drainage piping and extend to the open air outside of a building. Vent stacks are used as connection points for other vents, such as branch vents. A vent stack is a primary vent that accepts the connection of other vents and that vents an entire system.

Vent stacks run vertically and are sized a little differently from branch vents.

The basic procedure for sizing a vent stack is similar to that used with branch vents, but there are some differences. You must know the size of the soil stack (a pipe that conveys sewerage), the number of fixture-units carried by the soil stack, and the developed length of your vent stack. With this information and the regulations of your local plumbing code, you can size your vent stack. Let's work on an example.

Assume that your system has a soil stack with a diameter of 4 in. This stack is loaded with 43 fixture-units. Your vent stack will have a developed length of 50 ft. What size pipe will you have to use for your vent stack? When you look at the table in your code book, you see that a 2-in. pipe, used as a vent for the described soil stack, would allow a developed length of 35 ft. Your vent will have a developed length of 50 ft, so, you can rule out 2-in. pipe. In the column for 2½-in. pipe, you see a rating for up to 85 ft. Since your vent is going only 50 ft, you could use a 2½-in. vent. However, since 2½-in. pipe is not common, you would probably use a 3-in. pipe. This same sizing method is used when computing the size of stack vents.

STACK VENTS

Stack vents are really two pipes in one. The lower portion of the pipe is a soil pipe, and the upper portion is a vent. This is the type of primary vent most often found in residential plumbing. Stack vents are sized by using the same methods used in sizing vent stacks.

COMMON VENTS

Common vents are single vents that vent multiple traps and are allowed to be used only when the fixtures being served by the single vent are on the same floor level. Zone one requires the drainage of fixtures being vented with a common vent to enter the drainage system at the same level. Normally, not more than two traps can share a common vent, but there is an exception in zone three. Zone three allows you to vent the traps of up to three lavatories with a single common vent. Common vents are sized by using the same technique applied to individual vents.

WET VENTS

Wet vents are pipes that serve as a vent for one fixture and a drain for another. Wet vents, once you know how to use them, can save you a lot of money and time. By effectively using wet vents you can reduce the amount of pipe, fittings, and labor required to vent a bathroom group, or perhaps even two groups.

The sizing of wet vents is based on the number of fixture-units. The size of the pipe is determined by how many fixture-units it may be required to carry. A 3-in. wet vent can handle 12 fixture-units. A 2-in. wet vent is rated for four fixture-units, and a 1½-in. wet vent is allowed only one fixture-unit. It is acceptable to wet vent two bathroom groups, six fixtures, with a single vent, but the bathroom groups must be on the same floor level. Zone two makes provisions for wet venting bathrooms on different floor levels. Zone one takes a different approach to wet venting.

Zone two has some additional regulations that pertain to wet venting. The horizontal branch connecting to the drainage stack must enter at a level equal to, or below, the toilet drain. However, the branch may connect to the drainage at the bend of the toilet. When wet venting two bathroom groups, the wet vent must have a minimum diameter of 2 in. Kitchen sinks and washing machines may not be drained into a 2-in. combination waste and vent drain. Toilets and urinals are restricted on the use of vertical combined waste and vent systems.

As for zone two's allowance in wet venting on different levels, wet vents must have at least a 2-in. diameter. Toilets that are not located on the highest floor must be back-vented. If, however, the wet vent is connected directly to the toilet bend, with a 45° bend, the toilet being connected to it is not required to be back-vented, even if it is on a lower floor.

Zone one limits wet venting to vertical piping that is restricted to receiving the waste from only those fixtures with fixture-unit ratings of two, or less, and that vent no more than four fixtures. Wet vents must be one pipe size larger than normally required, but they must never be smaller than 2 in. in diameter.

CROWN VENTS

A crown vent extends upward from a trap or trap arm. Crown-vented traps are not allowed. When crown vents are used, they are normally used on trap arms, but even then they are not common. The vent must be on the trap arm, and it must be behind the trap by a distance equal to twice the pipe size. For example, on a 1½-in. trap, the crown vent would have to be on the trap arm 3 in. behind the trap.

VENTS FOR SUMPS AND SEWER PUMPS

When sumps and sewer pumps are used to store and remove sanitary waste, the sump must be vented. Zones one and two treat these vents about the same as vents installed on gravity systems.

When installing a pneumatic sewer ejector, run the sump vent to outside air, without tying it into the venting system for the standard sanitary plumbing system. This ruling on pneumatic pumps applies to all three zones. If your sump is equipped with a regular sewer pump, you may tie the vent from the sump back into the main venting system for the other sanitary plumbing.

Zone three has some additional rules. Sump vents may not be smaller than a 1¼-in. pipe. The size requirements for sump vents are determined by the discharge of the pump. For example, a sewer pump capable of producing 20 gal a minute could have its sump vented for an unlimited distance with a 1½-in. pipe. If the pump were capable of producing 60 gal/min, a 1½-in. pipe could not have a developed length of more than 75 ft.

In most cases, a 2-in. vent is used on sumps, and the distance allowed for developed length is not a problem.

However, if your pump will pump more than 100 gal/min, you had better take the time to do some math. Your code book will provide you with the factors you need to size your vent, and the sizing is easy. You simply look for the maximum discharge capacity of your pump and match it with a vent that allows the developed length you need.

SPECIFIC DATA

Specific data from your local plumbing code are required to size a vent system accurately. The tables provided here give you a serious sampling of the type of local information you need in order to create acceptable vent systems.

SIZING A VENT STACK FOR WET-VENTING IN ZONE TWO

Wet-Vented Fixtures	Stack Size Required (inches)
1 to 2 Bathtubs or showers	2
3 to 5 Bathtubs or showers	2½
6 to 9 Bathtubs or showers	3
10 to 16 Bathtubs or showers	4

SIZING A WET STACK VENT IN ZONE TWO

Pipe Size of Stack (inches)	Fixture-Unit Load on Stack	Maximum Length of Stack
2	4	30
3	24	50
4	50	100
6	100	300

VENT SIZING FOR ZONE THREE
(FOR USE WITH VENT STACKS AND STACK VENTS)

Drain Pipe Size (inches)	Fixture-Unit Load on Drain Pipe	Vent Pipe Size (inches)	Maximum Developed Length of Vent Pipe (feet)
1½	8	1¼	50
1½	8	1½	150
1½	10	1¼	30
1½	10	1½	100
2	12	1½	75
2	12	2	200
2	20	1½	50
2	20	2	150
3	10	1½	42
3	10	2	150
3	10	3	1040
3	21	1½	32
3	21	2	110
3	21	3	810
3	102	1½	25
3	102	2	86
3	102	3	620
4	43	2	35
4	43	3	250
4	43	4	980
4	540	2	21
4	540	3	150
4	540	4	580

**VENT SIZING FOR ZONE THREE
(FOR USE WITH INDIVIDUAL, BRANCH, AND CIRCUIT VENTS
FOR HORIZONTAL DRAIN PIPES)**

Drain Pipe Size (inches)	Drain Pipe Grade per Foot (inches)	Vent Pipe Size (inches)	Maximum Developed Length of Vent Pipe (feet)
1½	1/4	1¼	Unlimited
1½	1/4	1½	Unlimited
2	1/4	1¼	290
2	1/4	1½	Unlimited
3	1/4	1½	97
3	1/4	2	420
3	1/4	3	Unlimited
4	1/4	2	98
4	1/4	3	Unlimited
4	1/4	4	Unlimited

5

Sizing Potable Water Systems

Some common practices are used when sizing potable water systems for dwellings. Three-quarter-inch pipe is normally run to the water heater, and it is typically used as a main water-distribution pipe. When nearing the end of a run, when there are no more than two fixtures to connect to, the 3/4-in. pipe is reduced to 1/2-in. pipe. Most water services are of a 3/4-in. diameter, with those serving houses with numerous fixtures being a 1-in. pipe. This rule-of-thumb sizing works on almost any single-family residence.

The size of water suppliers to fixtures are required to meet minimum standards, which are derived from local code requirements. You simply find the fixture you are sizing the supply for, and check the column heading for the proper size. Most code requirements seem to agree that there is no definitive boilerplate formula for establishing potable water-pipe sizing. Code officers can require pipe sizing to be performed by a licensed engineer. In most major plumbing systems, the pipe sizing is done by a design professional.

Code books give examples of how a system might be sized, but the examples are not meant to be a code requirement. The code requires a water system to be sized properly. However, owing to the complexity of the process, the books do not set firm statistics for the process. Instead, they give parts of the puzzle in the form of some minimum standards, but it is up to a professional designer to come up with an approved system.

Most codes assign a fixture-unit value to common plumbing fixtures. To size by using the fixture-unit method, you must establish the number of fixture-units to be carried by the pipe. You must also know the working pressure of the water system. Most codes provide guidelines for these two pieces of information.

For our example, we have a house with the following fixtures: three toilets, three lavatories, one bathtub-shower combination, one shower, one dishwasher, one kitchen sink, one laundry hookup, and two sillcocks. The water pressure serving this house is 50 psi. There is a 1-in. water meter serving the house, and the water service is 60 ft in length. With this information and the guidelines provided by your local code, you can do a good job of sizing your potable water system.

The first step is to establish the total number of fixture-units on the system. The code regulations provide this information. Let's say that you have three toilets, and that that's nine fixture-units. The three lavatories add three fixture-units. The tub-shower combination counts as two fixture-units (the shower head over the bathtub doesn't count as an additional fixture). The shower has two fixture-units. The dishwasher adds two fixture-units and so does the kitchen sink. The laundry hookup counts as two fixture-units. Each sillcock is worth three fixture-units. This house has a total load of 28 fixture-units.

Now you have the first piece of your sizing puzzle in place. The next step is to determine what size pipe will allow your number of fixture-units. Refer to the tables and charts in your code book pertaining to pressure rating, pipe length, and fixture loads.

As mentioned, our subject house has a water pressure of 50 psi. First, find the proper water meter size. The one you are looking for is 1 in. Notice that a 1-in. meter and a 1-in. water service are capable of handling 60 fixture-units when the pipe runs only 40 ft. However, when the pipe length is stretched to 80 ft, the fixture load drops to 41. At 200 ft, the fixture rating is 25. What is it at 100 ft? At 100 ft, the allowable fixture load is 36.

Now, what does this tell us? Well, we know that the water service is 60 ft long. Once inside the house, how far is it to the most remote fixture? In this case, the farthest fixture is 40 ft from the water service entrance. This gives us a developed length of 100 ft: 60 ft for the water service and 40 ft for the interior pipe. Looking in our code book, we see that for 100 ft of pipe, under the conditions in this example, we are allowed 36 fixture-units. The house has only 28 fixture-units, so our pipe sizing is correct.

What would happen if the water meter were a 3/4-in. meter instead of a 1-in. meter? With a 3/4-in. meter and a 1-in. water service and main distribution pipe, we could have 33 fixture-units. This would still be a suitable arrangement, since we have only 28 fixture-units. Could we use a 3/4-in. water service and water distribution pipe with the 3/4-in. meter? No, we couldn't. With all sizes set at 3/4 in., the maximum number of fixture-units allowed is 17.

In this example, the piping is oversized. But if you want to be safe, use this procedure. If you are required to provide a riser diagram showing the minimum pipe sizing allowed, you will have to do a little more work. Once inside a building, water distribution pipes normally extend for some distance, supplying many fixtures with water. As the distribution pipe continues on its journey, it reduces the fixture load as it goes.

For example, assume that the distribution pipe serves a full bathroom group within 10 ft of the water service. Once this group is served with water, the fixture-unit load on the remainder of the water distribution piping is reduced by six fixture-units. Because the pipe serves other fixtures, the fixture-unit load continues to decrease. So it is feasible for the water distribution pipe to become smaller as it goes along.

Now, let's take our same sample house and see how we could use smaller pipe. Okay, we know we need a 1-in. water service. Once inside the foundation, the water service becomes the water distribution pipe. The water heater is located 5 ft from the cold-water distribution pipe. The 1-in.

pipe will extend over the water heater and supply it with cold water. Then there will be a hot-water distribution pipe originating at the water heater. Now you have two water distribution pipes to size.

When sizing the hot- and cold-water pipes, you can make adjustments for fixture-unit values on some fixtures. For example, a bathtub is rated as two fixture-units. Because the bathtub rating is inclusive of both hot and cold water, obviously the demand for just the cold water pipe is less than that shown in our table. For simplicity's sake, I will not break the fixture-units down into fractions or reduced amounts but will show you the example as if a bathtub required two fixture-units of hot water and two fixture-units of cold water. However, you could reduce the amounts listed in the table by about 25 percent to obtain the rating for each hot- and cold-water pipe. For example, the bathtub, when being sized for only cold water, could take on a value of 1½ fixture-units.

Now then, let's get on with the exercise. We are at the water heater. We ran a 1-in. cold-water pipe overhead and dropped a 3/4-in. pipe into the water heater. What size pipe do we bring up for the hot water? First, count the number of fixtures that use hot water, and assign them a fixture-unit value. All the fixtures use hot water except the toilets and sillcocks. The total count for hot-water fixture-units is lucky number thirteen. From the water heater, our most remote hot water fixture is 33 ft away.

What size pipe should we bring up from the water heater? By looking at the table in our code book, we find a distance and fixture-unit count that will work in this case. Look under the 40-ft column because our distance is less than 40 ft. The first fixture-unit number you see is nine; this won't work. The next number is 27, which will work because it is greater than the 13 fixture-units we need. Looking across the table, you see that the minimum pipe size to start with is a 3/4-in. pipe. Isn't it convenient that the water heater just happens to be sized for 3/4-in. pipe?

Okay, now we start our hot-water run with 3/4-in. pipe. As our hot water pipe goes along the 33-ft stretch, it provides water to various fixtures. When the total fixture count remaining to be served drops to fewer than nine fixture-units, we can reduce the pipe to 1/2-in. pipe. We can also run our fixture branches off of the main in 1/2-in. pipe. We can do this because the highest fixture-unit rating on any of our hot-water fixtures is two fixture-units. Even with a pipe run of 200 ft we can use 1/2-in. pipe for up to four fixture-units. Is this sizing starting to ring a bell? Remember the rule-of-thumb sizing I gave you earlier? These sizing examples are making the rule-of-thumb method ring true.

With the hot-water sizing done, let's look at the cold-water piping. We have fewer than 40 ft to our farthest cold-water fixture. We branch off near the water heater drop for a sillcock, and there is a full bathroom group within 7 ft of our water heater drop. The sillcock branch can be 1/2-in. pipe. The pipe going under the full bathroom group could probably be reduced to 3/4 in., but it would be best to run it as a 1-in. pipe. However, after serving the bathroom group and the sillcock, how many fixture-units are left? Only 19. We can now reduce to 3/4-in. pipe, and when the demand drops to below nine fixture-units we can reduce to 1/2-in. pipe. All our fixture branches can be run with 1/2-in. pipe.

This is one way to size a potable water system that works, without being driven crazy by computations. There may be some argument about the sizes I gave in these examples. The argument would be that some of the pipe is oversized, but as I said earlier, when in doubt, go bigger. In reality, the cold-water pipe in the last example could probably have been reduced to 3/4-in. pipe where the transition was made from water service to water distribution pipe. It could have almost certainly been reduced to 3/4 in. after the water heater drop. Local codes have their own interpretation of pipe sizing, but this method will normally serve you well.

REFERENCE TABLES

Reference tables supplied with your local plumbing code will dictate fixture units and criteria for pipe sizing. The tables and formulas in this chapter simulate the type of references provided in local plumbing codes. Once you become comfortable using these types of data, you can size potable water systems effectively.

SAMPLE PRESSURE AND PIPE TABLE FOR SIZING WATER PRESSURE RANGING FROM 46 TO 60 PSI

Size of Water Meter and Street Service (inches)	Size of Water Service and Distribution Pipes (inches)	Number of Fixture Units for Maximum Length of Water Pipe (feet)					
		40	60	80	100	150	200
3/4	1/2	9	8	7	6	5	4
3/4	3/4	27	23	19	17	14	11
3/4	1	44	40	36	33	28	23
1	1	60	47	41	36	30	25
1	1¼	102	87	76	67	52	44

EQUIVALENT FIXTURE-UNIT RATINGS

Fixture	Hot and Cold Water Combined
Toilet	3
Lavatory	1
Tub/shower combination	2
Shower	2
Dishwasher	2
Kitchen sink	2
Laundry hookup	2
Sillcock	3

COMMON MINIMUM SIZES OF PIPE FOR FIXTURE SUPPLY

Fixture	Minimum Size (inches)
Bathtub	1/2
Bidet	3/8
Shower	1/2
Toilet	3/8
Lavatory	3/8
Kitchen sink	1/2
Dishwasher	1/2
Laundry tub	1/2
Hose bibb	1/2

FIXTURE-UNIT RATINGS IN ZONE THREE

Fixture	Rating
Bathtub	2
Shower	2
Residential toilet	4
Lavatory	1
Kitchen sink	2
Dishwasher	2
Clothes washer	3
Laundry tub	2

6

Pipe Sizing for Storm Water

Pipe sizing for storm water intimidates some plumbers. As an instructor of both code and apprenticeship classes at a technical college, I witnessed a number of plumbers and soon-to-be plumbers cringe at the thought of sizing a storm-water system. Their fear was real but unfounded. The requirements for designing storm-water systems can require more math skills than other types of sizing exercises, but the principles are similar and the work is bearable. Knowledge and confidence are the keys to success, and this chapter should provide you with both.

SIZING A HORIZONTAL STORM DRAIN OR SEWER

The first step to take when sizing a storm drain or sewer is to establish your known criteria. How much pitch will your pipe have on it? In the example I am going to give you, my pipe will have a pitch of 1/4 in./ft. Because I know the pitch I will be using, this gives me a starting point and begins to take the edge off of an intimidating task.

What else do I know? Well, I know my system is going to be located in Portland, Maine. Portland's rainfall is rated at 2.4 in./hr. This rating assumes a 1-hr storm that is likely to occur only once every 100 years. I now have two of the factors I need to size my system.

77

I also know that the surface area my system will be required to drain is 15,000 ft^2; this includes the roof and parking area. I've now got three of the elements needed to get this job done. But how do I use the numbers? I must use them in conjunction with tables and charts in my local code book. When working with a standard table, such as those found in most code books, you must convert the information to suit your local conditions. For example, if a standardized table is based on 1 in./hr of rainfall and my location has 2.4 in./hr, I must convert the table. But this is not difficult; trust me.

When I want to convert a table based on a 1-in. rainfall to meet my local needs, all I have to do is divide the drainage area in the table by my rainfall amount. For example, if my standard table shows an area of 10,000 ft^2 requiring a 4-in. pipe, I can change the table by dividing my rainfall amount, 2.4, into the surface area of 10,000 ft^2.

If I divide 10,000 by 2.4, I get 4167. All of a sudden, I have solved the mystery of computing storm-water piping needs. With this simple conversion, I know that if my surface area is 4167 ft^2, I need a 4-in. pipe. But my surface area is 15,000 ft^2, so what size pipe do I need? Well, I know it will have to be larger than 4 in. So I look down my conversion table and find the appropriate surface area. My 15,000 ft^2 of surface area will require a storm-water drain with a diameter of 8 in.

Now, let's recap this exercise. To size a horizontal storm drain or sewer, decide what pitch you will put on the pipe. Next, determine what your area's rainfall is for a 1-hr. storm occurring each 100 yrs. If you live in a city, its rainfall amount may be listed in your code book. Using a standardized chart rated for 1 in./hr of rainfall, divide the surface area by a factor equal to your rainfall index. In my case it was 2.4. This division process converts a generic table into a customized table just for your area.

Once the math is done, look down the table for the surface area that most closely matches the area you have to drain. To be safe, go with a number slightly higher than your projected number. It is better to have a pipe sized one size too large than one size too small. When you have found the appropriate surface area, look across the table to see what size pipe you need.

SIZING RAIN LEADERS AND GUTTERS

When you are required to size rain leaders (or downspouts) and gutters, use the same procedures that you would use for storm drains—with one exception. Use a table, supplied in your code book, to size the vertical piping. Determine the amount of surface area your leader will drain, and use the appropriate table to establish your pipe size. The conversion factors are the same. Sizing gutters is essentially the same as sizing horizontal storm drains. The table provided in your code book is different, but the mechanics are the same.

ROOF DRAINS

Roof drains are often the starting point of a storm-water drainage system. As the name implies, roof drains are located on roofs. On most roofs, the roof drains are equipped with strainers that protrude upward at least 4 in. to catch leaves and other debris. Roof drains should be at least twice the size of the piping connected to them. All roofs that do not drain to hanging gutters are required to have roof drains. A minimum of two roof drains should be installed on roofs with a surface area of 10,000 ft^2 or less. If the surface area exceeds 10,000 ft^2, a minimum of four roof drains should be installed.

When a roof is used for purposes other than shelter, the roof drains may have a strainer that is flush with the roof's surface. Roof drains should obviously be sealed to prevent water from leaking around them. The size of the roof drain can be instrumental in the flow rates designed into a storm-water system. When a controlled flow from roof drains is wanted, the roof structure must be designed to accommodate the controlled flow.

If a combined storm-drain and sewer arrangement is approved, it must be sized properly. This requires converting fixture-unit loads into drainage surface area. For example, 256 fixture-units would be treated as 1000 ft^2 of surface area. Each additional fixture-unit in excess of 256 is assigned a value of 3.9 ft^2. In sizing for continuous flow, each gallon per minute (gpm) is rated as 96 ft^{2} of drainage area.

STORM-WATER PIPING

Storm-water piping requires the same amount of clean-outs, with the same frequency, as a sanitary system. Just as regular plumbing pipes must be protected, so must storm-water piping be. For example, if a downspout is in danger of being crushed by automobiles, you must install a guard to protect the downspout.

Storm-water systems and sanitary systems should not normally be combined. There may be some cities in which the two may be combined, but they are the exception rather than the rule. Areaway drains or floor drains must be trapped. When rain leaders and storm drains are allowed to connect to a sanitary sewer, it is required that they be trapped. The trap must be equal in size to the drain it serves. Traps must be accessible for cleaning the drainage piping. Storm-water piping may not be used for conveying sanitary drainage.

SUMP PUMPS

Sump pumps are used to remove water collected in building subdrains. These pumps must be placed in a sump, but the sump need not be covered with a gas-tight lid or be vented.

Many people are not sure what to do with the water pumped out of their basement by a sump pump. Do you pump it into your sewer? No, the discharge from a sump pump should not be pumped into a sanitary sewer. The water from the pump should be pumped to a storm-water drain or, in some cases, to a point on the property where it will not cause a problem.

All sump pump discharge pipes should be equipped with a check valve. The check valve prevents previously pumped water from running down the discharge pipe and refilling the sump, forcing the pump to pull double duty. When I speak of sump pumps, I am talking about pumps removing ground water, not waste or sewage.

Zone one has some additional requirements. Once storm-water piping extends at least 2 ft from a building, any approved material may be used. In zone one, the inlet area of a roof drain is generally required to be only 1½ times the size of the piping connected to the roof drain. However, when positioned on roofs used for purposes other than weather protection, roof drain openings must be sized to be twice as large as the drain connecting to them.

Zone one also provides tables for sizing purposes. When computing the drainage area, you must take into account the effect vertical walls have on a drainage area. For example, a vertical wall that reflects water onto the drainage area must be allowed for in computations of surface area. In the case of a single vertical wall, add half of the wall's total square footage to the surface area.

Two vertical walls adjacent to each other require you to add 35 percent of the combined square footage of the walls to the surface area.

If you have two walls of the same height that are opposite each other, no added space is needed. In this case, each wall protects the other one and does not allow extra water to collect on the roof area.

When you have two opposing walls with different heights, you must make an adjustment in the surface area. Take the square footage of the highest wall, as it extends above the other wall, and add half of its square footage to the surface area.

When you encounter three walls, use a combination of the preceding instructions to reach your goal. Four walls of equal height do not require an adjustment. If the walls are not of equal height, use the foregoing procedures to compute the surface area.

It would be helpful if all plumbing codes were the same, but they are not. The following information provides insight into how zone two varies from zones one and three. In zone two sump pits are required to have a minimum diameter of 18 in. Floor drains may not connect to drains intended solely for storm water. When computing surface area to be drained for vertical walls, such as walls enclosing a rooftop stairway, use half of the total square footage from the vertical wall surface that reflects water onto the drainage surface.

Some roof designs require a backup drainage system in case of emergencies. These roofs are generally those surrounded by vertical sections. If these vertical sections are capable of retaining water on the roof and if the primary drainage system fails, a secondary drainage system is required. In these cases, the secondary system must have independent piping and discharge locations. These special systems are sized by using different rainfall rates. The ratings are based on a 15-min rainfall. Otherwise, the 100-year conditions still apply.

For sizing a continuous flow, zone two requires a rating of 24 ft^2 of surface area to be given for every gpm generated. For regular sizing that is based on 4 in./hr of rain, 256 fixture-units equal 1000 ft^2 of surface area. Each additional fixture unit is rated at 3.9 in. If the rainfall rate varies, a conversion must be done.

To convert the fixture-unit ratings to a higher or lower rainfall, you must do some math. Take the square-foot rating assigned to fixture-units and multiply it by 4. For example, 256 fixture-units equal 1000 ft^2. Multiply 1000 by 4 and get 4000. Now divide 4000 by the rate of rainfall for 1 hr. Say, for example, that the hourly rainfall is 2 in.; the converted surface area would be 2000 ft^2.

A HORIZONTAL STORM-WATER SIZING TABLE[a]

Pipe Grade Per Foot (inches)	Pipe Size (inches)	Gallons per Minute (gpm)	Square Feet of Surface Area
1/4	3	48	4,640
1/4	4	110	10,600
1/4	6	314	18,880
1/4	8	677	65,200

[a]These figures are based on a rainfall with a maximum rate of 1 in./hr of rain, for a full hour and occurring once every 100 years.

APPROVED MATERIALS FOR STORM-WATER
DRAINAGE IN ZONE TWO[a]

For Underground Use	For Aboveground Use	For Storm Sewers
Cast iron	Galvanized steel	Cast iron
Coated aluminum	Black steel	Aluminum[b]
ABS[a] (acrylonitrile	Brass	ABS
butadiene styrene)	DWV copper or thicker	PVC
PVC[a] (polyvinyl chloride)	types of copper	Vitrified clay
Copper[a]	Cast iron	Concrete
	ABS	Asbestos-cement
Concrete[a]	PVC	
Asbestos-cement[a]	Aluminum	
Vitrified clay[a]	Lead	

[a]May be allowed for use, subject to local code authorities.
[b]Buried aluminum must be coated.

APPROVED MATERIALS FOR STORM-WATER
DRAINAGE IN ZONE ONE

Inside Buildings, Aboveground	Inside Buildings, Below Ground	Building Exteriors
Galvanized steel	Service-weight cast iron	Sheet metal with a
Wrought iron	Copper	minimum gauge of 26
Brass	ABS	
Copper	PVC	
Cast iron	Extra-strength vitrified clay	
ABS[a]		
PVC[a]		
Lead		

[a]ABS and PVC may not be used in buildings that have more than three floors above grade.

SOIL ABSORPTION RATINGS

	Slow Absorption	Medium Absorption	Rapid Absorption
Seconds required for water to drop 1 in.	5–30	3–5	0–3

LENGTH OF SUBSOIL DRAINAGE LINES

Number of People Served	Slow Absorption	Medium Absorption	Rapid Absorption
1–4	200	150	100
5–9	700	350	200
10–14	1000	500	340
15–20	1250	650	475

7

Approved Drainage and Venting Materials

Not all types of materials are approved for drainage use. Materials that are allowed for an aboveground drain may not be allowed for underground piping. The same can hold true for materials used to build a vent system. Some materials are approved for limited types of drainage piping. If you are going to design or install plumbing systems, you must know the types of materials that may be used and the restrictions on their use. This chapter will help you in this regard.

APPROVED MATERIALS FOR DRAIN, WASTE, AND VENT (DWV) PIPE

Acrylonitrile Butadiene Styrene (ABS)

Pipe made of ABS is black or sometimes a dark gray color. It is labeled as a DWV pipe when it is meant for DWV purposes. Pipe made of ABS is normally used as a DWV pipe instead of as a water pipe. The standard weight rating for common DWV pipe is schedule 40.

Pipe of ABS is easy to work with and may be used above or below ground. This material is joined with a solvent-weld cement and rarely leaks, even in less than desirable installation circumstances. It is extremely durable and can take hard abuse without breaking or cracking.

In zone one, the use of ABS is restricted to certain types of structures. It may not be used in buildings that have more than three habitable floors. The building may have a buried basement, where at least half of the exterior wall sections are at ground level or below. The basement may not be used as habitable space. So it is possible to use ABS in a four-story building, so long as the first story is buried in the ground, as stated, and not used as living space.

Coated Aluminum

Aluminum tubing is approved for aboveground use only. Aluminum tubing may not be allowed in zone one. Aluminum tubing is usually joined with mechanical joints and coated to prevent corrosive action. This material is, like most others, available in many sizes. The use of aluminum tubing has not become common for average plumbing installations in most regions.

Borosilicate Glass

Borosilicate glass pipe may be used above or below ground for DWV purposes in zone two. Underground use requires a heavy schedule of pipe. Zones one and three do not recognize this pipe as an approved material. However, as with all regulations, local authorities have the power to amend regulations to suit local requirements.

Brass

Brass pipe could be used as a DWV pipe, but it rarely is. The degree of difficulty in working with it is one reason it is not used more often. Zones two and three do not allow brass pipe to be used below grade for DWV purposes.

Cast Iron

Cast-iron pipe has long been a favored DWV pipe. Cast iron has been used for many years and provides good service for extended periods. The pipe is available in a hub-and-spigot style, the type used years ago, and in a hubless version. The hubless version is newer and is joined with mechanical joints that resemble a rubber coupling and that are surrounded and compressed by a stainless steel band.

The older, bell-and-spigot (or hub-and-spigot) type of cast iron is the type most often encountered during remodeling jobs. This type of cast iron was normally joined with the use of oakum and molten lead. However, today rubber adapters are available for creating joints with this type of pipe. These rubber adapters also allow plastic pipe to be mated to the cast iron.

Cast-iron pipe is frequently referred to as *soil pipe.* This nickname separates DWV cast iron from cast iron designed for use as a potable water pipe. Cast iron is available as a service-weight pipe and as an extra-heavy pipe. Service-weight cast iron is the most commonly used. Even though the cost of labor and material for installing cast iron is more than it is for schedule-40 plastic, cast iron still is in use, both above and below grade. It is sometimes used in multifamily dwellings and custom houses to deaden the sound of drainage as it passes down the pipe in walls adjacent to living space. If chemical or heat concerns are present, cast iron is often chosen over plastic pipe.

Copper

Copper pipe is made in a DWV rating. It is thin-walled and identified by a yellow marking. The pipe is a good DWV pipe, but it is expensive and time-consuming to install. Copper is not normally used in new installations unless extreme temperatures, such as those from a commercial dishwasher, warrant the use of a nonplastic pipe. Copper DWV is approved for use above and below ground in zones one and two. Zone three requires a minimum copper rating of type L for copper used underground for DWV purposes.

Galvanized Steel

Galvanized steel pipe keeps popping up as an approved material, but it is no longer a good choice for most plumbing jobs. Because galvanized steel pipe ages and rusts, the rough surface caused by the rust is prone to catching debris and creating pipe blockages. Another disadvantage to galvanized steel DWV pipe is the time it takes to install it. Galvanized steel pipe is not allowed for underground use in DWV systems. When used for DWV purposes, galvanized steel pipe should not be installed closer than 6 in. to the earth.

Lead

Lead pipe is still an approved material, but like galvanized steel pipe, it has little place in modern plumbing applications. Zone two does not allow the use of lead for DWV installations. Zone three limits the use of lead to above-grade installations.

Polyvinyl Chloride (PVC)

Pipe made of PVC is probably the leader in today's DWV pipe. This plastic pipe is white and is normally used in a rating of schedule 40. It uses a solvent weld joint and should be cleaned and primed before being glued together. This pipe becomes brittle in cold weather. If PVC is dropped on a hard surface while the pipe is cold, it is likely to crack or shatter. The cracks can go unnoticed until the pipe is installed and tested. It may be used above or below ground.

In zone one, the use of PVC is restricted to certain types of structures. It may not be used in buildings that have more than three habitable floors. The building may have a buried basement, of which at least half of the exterior wall sections are at ground level, or below. The basement may not be used as habitable space. So it is possible to use PVC in a four-story building, provided the first story is buried in the ground, as stated, and not used as living space.

SEWER PIPING

If underground piping will be used as a building sewer, one of the following types of pipes may be used:

ABS	PVC
Cast iron	Concrete
Vitrified clay	Asbestos cement

In zone three, bituminized-fiber pipe, type L copper pipe, and type K copper pipe may be used. Zone one tends to stick to the general guidelines, as given in the previous paragraphs.

If a building sewer is to be installed in the same trench that contains a water service, some of the pipes in the foregoing list may not be used. Standard procedure for pipe selection under these conditions calls for the use of a pipe approved for use inside a building. These types of pipes include ABS, PVC, and cast iron. Pipes that are more prone to breakage, such as a clay pipe, are not allowed unless special installation precautions are taken.

Sewers installed in unstable ground are also subject to modified rulings. Normally, any pipe approved for use underground, inside a building, will be approved for use with unstable ground. But the pipe must be well supported for its entire length.

Chemical wastes must be conveyed and vented by a system separated from the building's normal DWV system. The material requirements for chemical waste piping must be obtained from the local office of code enforcement.

FLANGES

Toilet flanges made from plastic must have a thickness of 1/4 in. Brass flanges may have a thickness of only 1/8 in. Flanges intended for caulking must have a thickness of 1/4 in., with a caulking depth of 2 in. The screws or bolts used to secure flanges to a floor must be brass. All flanges must be approved for use by the local authorities.

Zone two prohibits the use of offset flanges without prior approval. Zone three requires hard-lead flanges to weigh at least 25 oz and to be made from a lead alloy with no less than 7.75 percent antimony, by weight. Zone one requires flanges to have a diameter of about 7 in. In zone one, the combination of the flange and the pipe receiving it must provide about 1½ in. of space to accept the wax ring or sealing gasket.

FITTINGS

Fittings approved for *horizontal to vertical* changes in zone one include the following:

45° wye

60° wye

combination wye and eighth bend

sanitary tee

sanitary tapped-tee branches

Cross fittings, like double sanitary tees, cannot be used when they are of a short-sweep pattern. However, double sanitary tees can be used if the barrel of the tee is at least two pipe sizes larger than the largest inlet.

Fittings approved for *horizontal* changes in zone one are the 45° wye and the combination wye and eighth bend. Other fittings with similar sweeps may also be approved.

Fittings approved for *vertical to horizontal* changes in zone one are 45° branches and 60° branches and offsets. The latter must be installed in a true vertical position.

MATERIALS APPROVED FOR VENTS IN ZONE ONE

Aboveground	*Underground*
Cast iron	Cast iron
ABS[a]	ABS[a]
PVC[a]	PVC[a]
Copper	Copper
Galvanized steel	Brass
Lead	Lead
Brass	

[a]These materials may not be used with buildings having more than three floors above grade.

MATERIALS APPROVED FOR VENTS IN ZONE TWO

Aboveground	*Underground*
Cast iron	Cast iron
ABS	ABS
PVC	PVC
Copper	Copper
Galvanized steel	Aluminum
Lead	Borosilicate glass
Aluminum	
Borosilicate glass	
Brass	

MATERIALS APPROVED FOR VENTS IN ZONE THREE

Aboveground	*Underground*
Cast iron	Cast iron
ABS	ABS
PVC	PVC
Copper	Copper
Galvanized steel	
Lead	
Aluminum	
Brass	

8

Approved Materials for Water Systems

This chapter provides you with information on approved materials for water systems, including water service and water distribution systems. Many rules govern the use of various materials for these purposes, and the rules are explained here.

APPROVED MATERIALS FOR WATER SERVICE

Acrylonitrile Butadiene Styrene (ABS)

Pipe made from ABS is a plastic pipe that is normally used for drains and vents, but if properly rated, it can be used for water service. It must meet certain specifications for pressure-rated potable water use and be approved by the local authorities.

Asbestos Cement

Asbestos cement pipe has been used for municipal water mains in the past, but it is not used much today. It is, however, still an approved material.

Brass

Brass pipe is an approved material for water service piping, but it is rarely used. The complications of placing this metallic, threaded pipe below ground discourage its use. Brass pipe can also be used as a water distribution pipe.

Cast Iron

Cast-iron water pipe is approved for use in a water service, but it should not be used for individual water supplies.

Copper

Copper tubing is often referred to as copper pipe, but there is a difference. Copper pipe can be found with or without threads. Copper pipe is marked with a gray color code and is approved for water service use, but copper tubing is the copper most often used by plumbers.

Copper tubing can be purchased as soft copper, in rolls, or as rigid copper in lengths resembling pipe. Copper tubing, frequently called copper pipe in the trade, has long been used as the plumber's workhorse. It is approved for water service and comes in many different grades. The three types of copper normally used are type M, type L, and type K. The type-rating refers to the wall thickness of the copper. Type K is the thickest, type M is the thinnest, and type L is in the middle. Type L is generally considered the most logical choice for a water service. It is thicker than type M and is less expensive than type K.

Chlorinated Polyvinyl Chloride (CPVC)

Pipe made of CPVC is a white or cream-colored plastic pipe and is allowed for use in water services when it is rated for potable water. The plastic is fairly fragile, especially when cold, and it requires joints if the run of pipe exceeds 20 ft. It is advisable to avoid joints in underground water services when using CPVC.

Galvanized Steel

Galvanized steel pipe is an approved material for water services, but it is not a good choice. The pipe is joined with threaded fittings. Over time, this pipe will rust, and the rust can occur at the threaded areas, where the pipe walls are weakened, or inside the pipe. When the threaded areas rust, they can leak. When the interior of the pipe rusts, it can restrict the flow of water and reduce water pressure. Although this gray metal pipe is still available, it is rarely used in modern applications.

Polybutylene (PB)

Polybutylene may very well be the pipe of the future. It is an amazing product. Polybutylene is available in rolled coils and in straight lengths, and it is approved for water service. It is available as a water-service-only pipe and as a water-service/water-distribution pipe. If the pipe is blue, it is intended for water service only. If the pipe is gray, it can be used for water service or water distribution. Before PB pipe can be used for potable water, it must be tested by a recognized testing agency and approved by local authorities. Polybutylene is relatively inexpensive, very flexible, and a good choice for most water service applications.

Polyethylene (PE)

Polyethylene is a black, or sometimes bluish, plastic pipe that is frequently used for water services. It resists chemical reactions, as does polybutylene, and it is fairly flexible. This pipe is available in long coils, allowing it to be rolled out for great distances without joints.

This pipe may be one of the most common materials used for water services. However, it is not rated as a water distribution pipe. It is subject to crimping in tight turns, but it is a good pipe that will give years of satisfactory service.

Polyvinyl Chloride (PVC)

Pipe made of PVC is well known as a drain and vent pipe, but the PVC used for water services is not the same pipe. Both pipes are white, but the PVC used for water services is not the same as that used for drains and vents. The PVC used for water services must be rated for use with potable water.

Water pipe made of PVC is not acceptable as a water distribution pipe. It is not approved for hot-water use. Remember, when we talk about water distribution pipes, we assume that the building is supplied with both hot and cold water. If the building's only water distribution is cold water, some of the pipes, such as those made of PVC and PE, can be used as water distribution pipes.

APPROVED MATERIALS FOR WATER DISTRIBUTION

Brass

Brass pipe is suitable for water distribution, but it is not normally used in modern applications. Newer materials now available are easier to work with and usually provide longer service with fewer problems.

Copper

Both copper pipe and copper tubing are acceptable choices for water distribution. Copper tubing is by far the more common choice. If water has an unusually high acidic content, copper can be subject to corrosion and pinhole leaks. If acidic water is suspected, a thick-walled copper pipe or a plastic pipe should be considered over the use of thin-walled copper.

Zones two and three allow types M, L, and K to be used above and below ground. But zone one does not allow the use of type M copper when the copper is installed underground within a building.

Galvanized Steel

Galvanized steel pipe remains in the approved category, but it is hardly ever used in new plumbing systems. The characteristics of galvanized pipe remove it from the competition. It is difficult to work with and is subject to rust-related problems. The rust can cause leaks and reduced water pressure and volume.

Polybutylene (PB)

Polybutylene is edging copper out of the picture in many locations. With its ease of installation, resistance to splitting during freezing conditions, and low cost, polybutylene is a strong competitor. Add to this the fact that PB resists chemical reactions, and you have yet another advantage over copper. Many old-school plumbers are dubious about the gray, plastic pipe, but it is making a good reputation for itself.

FITTINGS

Fittings that are made from cast iron, copper, plastic, and steel and other types of iron are all approved for use in their proper place. Generally speaking, fittings must either be made from the same material as the pipe they are being used with or be compatible with the pipe.

NIPPLES

Manufactured pipe nipples are normally made from brass or steel. They range in length from 1/8 in. to 12 in. Nipples must meet certain standards, but they should be rated and approved before you obtain them.

VALVES

Valves have to meet some standards, but most of the decisions for the use of valves will come from the local office of code enforcement. Valves, like fittings, must be either of the same material as the pipe they are being used with or compatible with the pipe. Size and construction requirements will be stipulated by local jurisdictions.

9

Materials for
Storm-Water Systems

Materials used for storm-water systems must conform to the
requirements of your local plumbing code. Your local code
book contains the information you need, but this chapter
serves as a handy, quick reference for your needs in selecting
storm-water materials. Keep in mind, however, that local
code requirements can vary a great deal, and it is always
best to confirm your thoughts with a current code book.

MATERIALS FOR STORM DRAINAGE

Inside Storm Drainage
Materials commonly approved in zone three for interior
storm drainage include the following:

ABS	PVC
Type DWV copper	Type M copper
Type L copper	Type K copper
Asbestos cement	Bituminized fiber
Cast iron	Concrete
Vitrified clay	Aluminum, coated
Brass	Lead
Galvanized steel	

Zone two's approved pipe materials for interior storm drainage include the following:

ABS	PVC
Type DWV copper	Type M copper
Type L copper	Type K copper
Asbestos cement	Cast iron
Concrete	Aluminum, coated
Galvanized steel	Black steel
Brass	Lead

Zone one follows its standard pipe approvals for storm-water piping.

Storm-Drainage Sewers

Storm-drainage sewers are a little different from inside storm-drainage systems. If you are installing a storm-drainage sewer, use one of the following types of pipes:

Zone One

Follow the basic guidelines for approved piping.

Zone Two

Cast iron	Asbestos cement
Vitrified clay	Concrete
ABS	PVC
Aluminum, coated	

Zone Three

Cast iron	Asbestos cement
Vitrified clay	Concrete
ABS	PVC
Bituminized fiber	Type L copper
Type K copper	Type M copper
Type DWV copper	

MATERIALS FOR SUBSOIL DRAINS

Subsoil drains are designed to collect and drain water entering the soil. They are frequently slotted pipes and can be made from any of the following materials in most cases:

Asbestos cement	Bituminized fiber
Vitrified clay	PVC
Cast iron	Polyethylene

Zone two does not allow bituminized fiber pipe to be used as a subsoil drain. It may not allow cast iron or some plastics. Zone one sticks to its normal pipe approvals.

10

The Basics of a
Potable Water System

This chapter covers the basics of a potable water system. Not all the rules and regulations are detailed here, but the ones you should need most often are referenced. Become acquainted with this material to save yourself time and trouble out on your next job.

WATER SERVICE

The main pipe delivering water to a potable water system is a water service. A water service pipe must have a diameter of at least 3/4 in. The pipe must be sized according to code requirements to provide adequate water volume and pressure to the fixtures.

Ideally, a water service pipe should be run from the primary water source to the building in a private trench. By private trench, I mean a trench not used for any purpose except for the water service. However, it is normally allowable to place the water service pipe in the same trench used by a sewer or building drain when specific installation requirements are followed. The water service pipe must be separated from the drainage pipe. The bottom of the water service pipe may not be closer than 12 in. to the drainage pipe at any point.

A shelf must be made in the trench to support the water service pipe. The shelf must be made solid and stable and be at least 12 in. above the drainage pipe. It is not acceptable to locate a water service pipe in an area where pollution is probable, and you should never run it through, above, or under a waste disposal system, such as a septic field.

If a water service is installed in an area subject to flooding, the pipe must be protected against flooding. Water services must also be protected against freezing. The depth of the water service depends on the climate of the location. Check with your local code officer to see how deep a water service pipe must be buried in your area. Take care when backfilling a water service trench. The backfill material must be free of objects, like sharp rocks, that may damage the pipe.

When a water service enters a building through or under the foundation, the pipe must be protected by a sleeve, which is usually a pipe with a diameter at least twice that of the water service. Once through the foundation, the water service may need to be converted to an acceptable water distribution pipe. Some materials approved for water service piping are not approved for interior water distribution.

If a water service pipe is not an acceptable water distribution material, it must be converted to an approved material, generally within the first 5 ft of its entry into the building. Once inside a building, the maze of hot and cold water pipes is referred to as water distribution pipes. Let's see what you need to know about water distribution systems.

FIXTURE SUPPLIES

Fixture supplies are the tubes or pipes that rise from the fixture branch, the pipe coming out of the wall or floor, to the fixture. In zone three, a fixture supply may not have a length of more than 30 in. The required minimum sizing for a fixture supply is determined by the type of fixture being supplied with water.

PRESSURE-REDUCING VALVES

Pressure-reducing valves are required to be installed on water systems when the water pressure coming to the water distribution pipes is in excess of 80 lb/in.^{2} The only time this regulation is generally waived is when the water service is bringing water to a device requiring high pressure.

PRESSURIZED WATER TANKS

Pressurized water tanks are the type most commonly encountered in modern plumbing. They are the type used with well systems. All pressurized water tanks should be equipped with a vacuum breaker. The vacuum breaker is installed on top of the tank and should be at least 1/2 in. in diameter. Vacuum breakers should be rated for proper operation up to maximum temperatures of 200°F.

It is also necessary to equip the tanks with pressure-relief valves, which must be installed on the supply pipe that feeds the tank or on the tank itself. The safety relief valve discharges when pressure builds to a point that could endanger the integrity of the pressure tank. The valve's discharge must be carried by gravity to a safe and approved disposal location. The piping carrying the discharge of a relief valve may not be connected directly to the sanitary drainage system.

SUPPORTING PIPES

The methods used for supporting pipes in potable water systems are regulated by the plumbing code. There are requirements for the types of materials you may use and how they may be used. One concern with the type of straphangers used is their compatibility with the pipe they are supporting. You must use a hanger that will not have a detrimental effect on your piping. For example, you may not use galvanized steel straphangers to support copper pipe. As a rule of thumb, the hangers used to support a pipe should be made from the same material as the pipe being supported. For example, copper pipe should be hung with copper hangers. This eliminates the risk of a corrosive action between two types of materials. If you use a plastic or plastic-coated hanger, you may use it with all types of pipe.

The hangers used to support pipe must be capable of supporting the pipe at all times. They must be attached to the pipe and to the member holding the hanger in a satisfactory manner. For example, it would not be acceptable to wrap a piece of wire around a pipe and then wrap the wire around the bridging between two floor joists. Hangers should be securely attached to the members supporting them. For example, a hanger should be attached to the pipe and then nailed to a floor joist. The nails used to hold a hanger in place should be made of the same material as the hanger if corrosive action is a possibility.

Both horizontal and vertical pipes require support. The intervals between supports will vary, depending upon the type of pipe being used and whether it is installed vertically or horizontally. The following examples will show you how often you must support the various types of pipes when they are hung horizontally. These examples give the maximum distances allowed between supports in zone three:

ABS—4 ft	Cast iron—5 ft
Galvanized steel—12 ft	PVC—4 ft
Copper—6 ft	Brass—10 ft
CPVC—3 ft	PB—32 in.

When these same types of pipes are installed vertically in zone three, they must be supported at no less than the following intervals:

ABS—4 ft	Cast iron—15 ft
Galvanized steel—15 ft	PVC—4 ft
Copper—10 ft	CPVC—3 ft
Brass—10 ft	PB—4 ft

WATER CONSERVATION

Water conservation continues to grow as a major concern. When setting the flow rates for various fixtures, water conservation is a factor. The flow rates of many fixtures must be limited to no more than 3 gal/min (gpm). In zone three, these fixtures include the following:

Showers	Lavatories
Kitchen sinks	Other sinks

The rating of 3 gpm is based on a water pressure of 80 psi.

When installed in public facilities, lavatories must be equipped with faucets producing no more than 1/2 gpm. If the lavatory is equipped with a self-closing faucet, it may produce up to 1/4 gpm per use. Toilets are restricted to a use of no more than 4 gal of water, and urinals must not exceed 1½ gal of water with each use.

ANTISCALD PRECAUTIONS

It is easy for the very young or the elderly to receive serious burns from plumbing fixtures. In an attempt to reduce accidental burns, it is required that mixed water to gang showers be controlled by thermostatic means or by pressure-balanced valves. All showers, except those in residential dwellings in zones one and two, must be equipped with pressure-balanced valves or thermostatic controls. These temperature-control valves may not allow water with a temperature of more than 120°F to enter the bathing unit. In zone three, the maximum water temperature is 110°F. Zone three requires these safety valves on all showers.

VALVE REGULATIONS

Gate valves and ball valves are examples of full-open valves, as required under valve regulations. The valves do not depend on rubber washers, and when they are opened to their maximum capacity, there is a full flow through the pipe. Many locations along the water distribution installation require the installation of full-open valves. Zone one requires these types of valves in the following locations:

On the water service, before and after the water meter
On each water service for each building served
On discharge pipes of water-supply tanks, near the tank
On the supply pipe to every water heater, near the heater
On the main supply pipe to each dwelling

In zone three, the locations for full-open valves are as follows:

On the water service pipe, near the source connections

On the main water distribution pipe, near the water service

On water supplies to water heaters

On water supplies to pressurized tanks, such as tanks for well systems

On the building side of every water meter

Zone two requires full-open valves to be used in all water distribution locations, except as cutoffs for individual fixtures in the immediate area of the fixtures. Other local regulations may apply to specific building uses; check with your local code officer to confirm where full-open valves may be required in your system. All valves must be installed so that they are accessible.

CUTOFF VALVES

Cutoff valves do not have to be full-open valves. Stop-and-waste valves are an example of cutoff valves that are not full-open valves. Every sillcock must be equipped with an individual cutoff valve. Appliances and mechanical equipment with water supplies are required to have cutoff valves installed in the service piping. Generally, with only a few exceptions, cutoffs are required on all plumbing fixtures. Check with your local code officer for fixtures not requiring cutoff valves. All valves installed must be accessible.

SUPPORT INTERVALS FOR WATER PIPE IN ZONE ONE

Type of Pipe (inches)	Vertical Support Interval	Horizontal Support Interval (feet)
Threaded pipe (3/4 and smaller)	Every other story	10
Threaded pipe (1 and larger)	Every other story	12
Copper tube (1½ and smaller)	Every story, not to exceed 10 ft	6
Copper tube (2 and larger)	Every story, not to exceed 10 ft	10
Plastic pipe	Not mentioned	4

SUPPORT INTERVALS FOR WATER PIPE IN ZONE TWO

Type of Pipe (inches)	Vertical Support Interval (feet)	Horizontal Support Interval (feet)
Threaded pipe	30	12
Copper tube (1¼ and smaller)	4	6
Copper tube (1½)	Every story	6
Copper tube (larger than 1½)	Every story	10
Plastic pipe (2 and larger)	Every story	4
Plastic pipe (1½ and smaller)	4	4

HOT-WATER INSTALLATIONS

When hot-water pipe is installed, it is often expected to maintain the temperature of the water for a distance of up to 100 ft from the fixture it serves. If the distance between the hot-water source and the fixture being served is more than 100 ft, a recirculating system is frequently required. When a recirculating system is not appropriate, other means can be used to maintain water temperature. These means can include insulation or heating tapes. Check with your local code officer for approved alternates to a recirculating system if necessary.

If a circulator pump is used on a recirculating line, the pump must be equipped with a cutoff switch, which may operate manually or automatically.

SOME HARD-LINE FACTS

- All devices used to treat or convey potable water must be protected against contamination.

- It is not acceptable to install stop-and-waste valves underground.

- If there are two water systems in a building, one potable, one nonpotable, the piping for each system must be marked clearly. The marking can be in the form of a suspended metal tag or a color code. Zone two requires the pipe to be color-coded and tagged. Nonpotable water piping should not be concealed.

- Hazardous materials, such as chemicals, may not be put in a potable water system.

- Piping that has been used for a purpose other than conveying potable water may not be used as a potable water pipe.

- Water used for any purpose should not be returned to the potable water supply but transported to a drainage system.

11

Plumbing Fixtures

No plumbing system is complete without fixtures, and fixtures can account for much of your need for knowledge. A lot of the rules that deal with fixtures are simple, but some of them are not so easy to grasp. Deciding when to install fixtures for handicapped people may be a problem for you, for example. Knowing how many toilets are required in the remodeling of a restaurant might throw you a curve. This chapter clears up a lot of potential confusion surrounding fixtures.

When you are planning fixtures for a single-family residence, you must include certain fixtures. If you choose to install more than the minimum, that's fine, but you must install the minimum number of required fixtures. The minimum number and types of fixtures for a single-family dwelling are as follows:

One toilet   One lavatory

One bathing unit   One kitchen sink

One hookup for a clothes washer

MULTIFAMILY BUILDINGS

The minimum requirements for a multifamily building are the same as those for a single-family dwelling, but the requirements are that each dwelling unit in the multifamily building must be equipped with the minimum number of fixtures. There is one exception—the laundry hookup. In a multifamily building, laundry hookups are not required in each dwelling unit. In zone three, it is required that a laundry hookup be installed for common use when the number of dwelling units is 20 or more. For each interval of 20 units, you must install a laundry hookup.

For example, in a building with 40 apartments, you would have to provide two laundry hookups. If the building had 60 units, you would need three hookups. In zone one, the dwelling-unit interval is 10 rental units. Zone two requires one hookup for every 12 rental units, but no fewer than two hookups for buildings with at least 15 units.

PLACES OF PUBLIC ASSEMBLY

When you get into businesses and places of public assembly, such a nightclubs and restaurants, the ratings are based on the number of people likely to use the facilities. The minimum requirements for such places in zone three is as follows:

- Toilets—one for every 40 people
- Lavatories—one for every 75 people
- Service sinks—one
- Drinking fountains—one for every 500 people
- Bathing units—none

DAY-CARE FACILITIES

The minimum number of fixtures for a day-care facility in zone three is as follows:

- Toilets—one for every 15 people
- Lavatories—one for every 15 people
- Bathing units—one for every 15 people
- Service sinks—one
- Drinking fountains—one for every 100 people

In contrast, zone two requires only the installation of toilets and lavatories in day-care facilities. The ratings for these two fixtures are the same as in zone three, but the other fixtures required by zone three are not required in zone two. This type of rating system can be found in your local code and will cover all the normal types of building uses.

In many cases, facilities must be provided in separate bathrooms to accommodate each sex. The number of required fixtures must be divided equally between the two sexes, unless there is cause and approval for a different appropriation.

Some types of buildings do not require separate facilities. For example, zone three does not require residential properties and small businesses, where fewer than 15 employees work or where fewer than 15 people are allowed in the building at the same time, to have separate facilities.

Zone two does not require separate facilities in offices with less than 1200 ft^2 of floor space, retail stores with less than 1500 ft^2, restaurants with less than 500 ft^2, self-serve laundries with less than 1400 ft^2, and hair salons with less than 900 ft^2.

EMPLOYEE AND CUSTOMER FACILITIES

Some special regulations pertain to employee and customer facilities. For employees, toilet facilities must be available to employees within a reasonable distance and with relative ease of access. For example, zone three requires these facilities to be in the immediate work area; the distance an employee is required to walk to the facilities may not exceed 500 ft. The facilities must be located in a manner so that employees do not have to negotiate more than one set of stairs for access to the facilities. There are some exceptions to these regulations, but in general these are the rules.

It is expected that restaurants, stores, and places of public assembly will provide toilet facilities. In zone three, this requirement is based on buildings capable of holding 150 or more people. Buildings in zone three with an occupancy rating of fewer than 150 people are not required to provide toilet facilities unless the building serves food or beverages. When facilities are required, they may be placed in individual buildings. In a shopping mall, for example, in a common area they must be placed not more than 500 ft from any store or tenant space and placed so that customers will not have to use more than one set of stairs to reach them.

Zone two uses a square-footage method to determine minimum requirements in public places. For example, retail stores are rated as having an occupancy load of one person for every 200 ft^2 of floor space. Such an occupancy load is required to have separate facilities when the store's square footage exceeds 1500 ft^2. A minimum of one toilet is required for each facility when the occupancy load is up to 35 people. One lavatory for up to 15 people is required in each facility. A drinking fountain is required for occupancy loads up to 100 people.

REQUIREMENTS FOR THE HANDICAPPED

Handicap fixtures are not cheap; you cannot afford to overlook them when bidding a job. Plumbing codes normally require specific minimum requirements for handicap-accessible fixtures in certain circumstances. It is your responsibility to know when handicap facilities are required. There are also special regulations pertaining to the installation of handicap fixtures.

When you deal with handicap plumbing, you must be guided by both the local plumbing code and the local building code. These two codes together establish the minimum requirements for handicap plumbing facilities. When you work in the field of handicap plumbing, you must follow a different set of rules. Handicap plumbing is a code unto itself.

Most buildings frequented by the public are required to have handicap-accessible plumbing fixtures. The following examples are based on zone three requirements. Zones one and two do not go into as much detail on handicap requirements in their plumbing codes.

Single-family houses and most residential multifamily dwellings are exempt from handicap requirements. A rule of thumb for most public buildings is the inclusion of one toilet and one lavatory for handicap use.

Hotels, motels, inns, and the like are required to provide a toilet, lavatory, bathing unit, and kitchen sink (where applicable) for handicap use. Drinking fountains may also be required. This provision will depend on the local plumbing and building codes. If plumbing a gang-shower arrangement, such as in a school gym, at least one of the shower units must be handicap-accessible. Door sizes and other building code requirements must be observed when dealing with handicap facilities. There are local exceptions to these rules, so check with your local code officers for current local regulations.

Toilet Facilities

When you think of a handicap toilet, you probably think of a toilet that sits high off the floor. But do you think of the grab bars and partition dimensions required around the toilet? Some plumbers don't, but they should. The door to a privacy stall for a handicap toilet must provide a minimum of 32 in. of clear space.

The distance between the front of the toilet and the closed door must be at least 48 in. It is mandatory that the door open outward, away from the toilet. Think about it. How could a person in a wheelchair close the door if the door opened inward? These facts may not seem like your problem, but if your inspection doesn't pass, you don't get paid.

The width of a compartment for handicap toilets must be a minimum of 5 ft. The length of the privacy stall must be at least 56 in. for wall-mounted toilets and 59 in. for floor-mounted models. Unlike regular toilets, which require a rough-in of 15 in. to the center of the drain from a sidewall, handicap toilets require the rough-in to be at least 18 in. off the sidewall.

Then there are the required grab bars. Sure, you may know that grab bars are required, but do you know the mounting requirements for them? Two bars are required for each handicap toilet. One bar should be mounted on the back wall and the other on the sidewall. The bar mounted on the back wall must be at least 3 ft long. The first mounting bracket of the bar must be mounted no more than 6 in.

from the sidewall. The bar must extend at least 24 in. past the center of the toilet's drain.

The bar mounted on the sidewall must be at least 42 in. long. It should be mounted level, with the first mounting bracket located no more than 1 ft from the back wall. The bar must be mounted on the sidewall that is closest to the toilet and must extend to a point at least 54 in. from the back wall. If you do your math, you will see that a 42-in. bar pushes the limits on both ends. A longer bar is more likely to meet the minimum requirements.

When a lavatory is to be installed in the same compartment as the toilet, the lavatory must be installed on the back wall. The lavatory must be installed so that its closest point to the toilet is no less than 18 in. from the center of the toilet's drain. When a privacy stall of this size and design is not suitable, there is an option, as follows.

Another way to size the compartment that is to house both a handicap toilet and lavatory is available. There may be times when space restraints do not allow a stall with a width of 5 ft. In these cases, you may position the fixture differently and use a stall with a width of only 3 ft. In these situations, the width of the privacy stall may not exceed 4 ft.

The depth of the compartment must be at least 66 in. when wall-mounted toilets are used. The depth extends to a minimum of 69 in. when a floor-mounted toilet is used. The toilet requires a minimum distance from sidewalls of 18 in. to the center of the toilet drain. If the compartment is more than 3 ft wide, grab bars are required, with the same installation methods as described before.

If the stall is made at the minimum width of 3 ft, grab bars with a minimum length of 42 in. are required on each side of the toilet. The bars must be mounted no more than 1 ft from the back wall, and they must extend a minimum of 54 in. from the back wall. If a privacy stall is not used, the sidewall clearances and the grab bar requirements are the same as given in these two examples. To determine which set of rules to use, you must assess the shape of the room when no stall is present.

If the room is laid out in a fashion like that of the first example, use the guidelines for grab bars given there. If the room meets the description of the second example, use the specifications there. In both cases, the door to the room may not swing into the toilet area.

FIXTURES FOR THE HANDICAPPED

Toilets

Toilets for the handicapped have a normal appearance, but they sit higher above the floor than a standard toilet. A handicap toilet rises to a height of between 16 and 20 in. off the finished floor, 18 in. being a common height. There are many choices in toilet style, including the following:

Siphon jet   Siphon vortex

Siphon wash   Reverse trap

Blowout

Sinks and Lavatories

Visually, handicap sinks, lavatories, and faucets may appear to be standard fixtures, but their method of installation is regulated and the faucets are often unlike a standard faucet. Handicap sinks and lavatories must be positioned to allow a person in a wheelchair to use them easily.

The clearance requirements for a lavatory are numerous. There must be at least 30 in. of clearance in front of the lavatory. This clearance must extend 30 in. from the front edge of the lavatory, or countertop, whichever protrudes the farthest, and to the sides. If you can set a square box with a 30×30-in. dimension in front of the lavatory or countertop, you have adequate clearance for the first requirement. This applies to kitchen sinks and lavatories.

The next requirement calls for the top of the lavatory to be no more than 35 in. from the finished floor. For a kitchen sink, the maximum height is 34 in. Then there is knee clearance to consider. The minimum allowable knee clearance requires 29 in. in height and 8 in. in depth, measured from the face of the lavatory or kitchen sink. Toe clearance is another issue. A space 9 in. high and 9 in. deep is required as a minimum for toe space. The last requirement deals with hot-water pipes. Any exposed hot-water pipes must be insulated, or shielded, in a way to prevent users of the fixture from being burned.

Sink and Lavatory Faucets
The faucets must be located no more than 25 in. from the front edge of the lavatory or counter, whichever is closer to the user. The faucets can be operated by wing handles, single handles, blade handles, or push buttons, but the operational force required by the user must not be more than 5 lb.

Bathing Units
Like any other fixtures, handicap bathtubs and showers must meet the approved requirements, but they are also required to have special features and to be installed by special methods. The special features and installation methods are required under the code for approved handicap fixtures. The clear space in front of a bathing unit is required to be a minimum of 1440 in.^{2}, which is achieved by leaving an open space of 30 in. in front of the unit and 48 in. on each side. If the bathing unit is not accessible from the side, the minimum clearance is increased to an area with a dimension of 48 in. by 48 in.

Handicap bathtubs must be installed with seats and grab bars. A grab bar for handicap use must have a diameter of at least 1¼ in. The diameter may not exceed 1½ in. All handicap grab bars are designed to be installed 1½ in. from walls. The design and strength for these bars are set forth in the building codes.

The seat may be an integral part of the bathtub, or it may be a removable, after-market seat. The grab bars must be at least 2 ft long. Two grab bars are to be mounted on the back wall, one over the other, running horizontally. The lowest grab bar must be mounted 9 in. above the flood-level rim of the tub. The top grab bar must be mounted a minimum of 33 in., but no more than 36 in., above the finished floor. The grab bars should be mounted near the seat of the bathing unit.

Additional grab bars are required at each end of the tub. They must be mounted horizontally and at the same height as the highest grab bar on the back wall. The bar over the faucet must be at least 2 ft long. The bar on the other end of the tub may be as short as 1 ft.

The faucets in these bathing units must be located below the grab bars. The faucets used with a handicap bathtub must be able to operate with a maximum force of 5 lb. A personal, hand-held shower is required in all handicap bathtubs. The hose for the hand-held shower must be at least 5 ft long.

Two types of showers are normally used for handicap purposes. The first type of shower allows the user to leave a wheelchair and shower while sitting on a seat. The other style of shower stall is meant for the user to roll a wheelchair into the stall and shower while seated in the wheelchair.

If the shower is intended to be used with a shower seat, its dimensions should form a square, with 3 ft of clearance from the seat. The seat should be no more than 16 in. wide, mounted along the sidewall, and running the full length of the shower. The height of the seat should be between 17 and 19 in. above the finished floor.

Two grab bars should be installed in the shower. They should be located between 33 and 36 in. above the finished floor. The bars are intended to be mounted in an L pattern. One bar should be 36 in. long and run the length of the seat, mounted horizontally. The other bar should be installed on the sidewall of the shower and should be at least 18 in. long.

The faucet for the seat type of shower must be mounted on the wall across from the seat. The faucet must be at least 38 in., but not more than 48 in., above the finished floor.

Drinking Units

The distribution of water from a watercooler or drinking fountain must occur at a maximum height of 36 in. above the finished floor. The outlet for drinking water must be located at the front of the unit, and the water must flow upward for a minimum distance of 4 in. Levers or buttons to control the operation of the drinking unit may be mounted on front of the unit or on the side, near the front.

Clearance requirements call for an open space of 30 in. in front of the unit and 48 in. to the sides. Knee and toe clearances are the same as required for sinks and lavatories. If the unit is made so that the drinking spout extends beyond the basic body of the unit, the width clearance may be reduced from 48 in. to 30 in., so long as knee and toe requirements are met.

INSTALLATION REGULATIONS FOR STANDARD FIXTURES

Standard fixtures must be installed according to local code regulations. There are space limitations, clearance requirements, and predetermined approved methods for installing standard plumbing fixtures. First, let's look at the space and clearance requirements for common fixtures.

Standard Fixture Placement

Toilets and bidets require a minimum distance of 15 in. from the center of the fixture's drain to the nearest sidewall. These fixtures must have at least 15 in. of clear space between the center of their drains and any obstruction, such as a wall, cabinet, or other fixture. With this rule in mind, a toilet or bidet must be centered in a space of at least 30 in. Zone one further requires that there be a minimum of 18 in. of clear space in front of these fixtures and that when toilets are placed in privacy stalls, the stalls must be at least 30 in. wide and 60 in. deep.

Zones one and two require urinals to be installed with a minimum clear distance of 12 in. from the center of their drains to the nearest obstacle on each side. When urinals are installed side by side in zones one and two, the distance between the centers of their drains must be at least 24 in.

Zone three requires urinals to have minimum sidewall clearances of at least 15 in. In zone three, the center-to-center distance is a minimum of 30 in. Urinals in zone three must also have a minimum clearance of 18 in. in front of them.

These fixtures, as with all fixtures, must be installed level and with good workmanship. The fixture should normally be set with an equal distance from walls to avoid a crooked or cocked installation. All fixtures should be designed and installed with proper cleaning in mind. Bathtubs, showers, vanities, and lavatories should be placed in a manner to avoid violating the clearance requirements for toilets, urinals, and bidets.

INSTALLATION OF TYPICAL RESIDENTIAL FIXTURES

Installation of typical residential fixtures includes everything from hose bibbs to bidets. This chapter takes each fixture that could be considered a typical residential fixture and tells you about how it must be installed.

The typical plumbing fixture has water coming into it and water going out of it. The incoming water lines must be protected against freezing and back-siphonage. Freeze protection is usually accomplished with the placement of the piping. In cold climates it is advisable to avoid putting pipes in outside walls. Insulation is often applied to water lines to reduce the risk of freezing. Back-siphonage is typically avoided with the use of air gaps and back-flow preventers.

Some fixtures, such as lavatories and bathtubs, are equipped with overflow routes. These overflow paths must be designed and installed to prevent water from remaining in the overflow after the fixture is drained. They must also be installed so that back-siphonage cannot occur. This normally means nothing more than having the faucet installed so that it is not submerged in water if the fixture floods. By keeping the faucet spout above the high-water mark, you have created an air gap. The path of a fixture's overflow must carry the overflowing water into the trap of the fixture by having the overflow path integrated with the same pipe that drains the fixture.

Bathtubs must be equipped with waste and overflow openings. Zone one and zone three require these openings to have a minimum diameter of 1½ in. The method for blocking the waste opening must be approved. Common methods for holding water in a tub include the following:

Plunger-style stoppers   Rubber stoppers
Lift and turn stoppers   Push and pull stoppers

Some fixtures, such as hand-held showers, pose special problems. Because the shower is on a long hose, it can be dropped into a bathtub full of water. If a vacuum is formed in the water pipe while the shower head is submerged, the unsanitary water from the bathtub can be pulled back into the potable water supply. This is avoided with the use of an approved back-flow preventer.

When a drainage connection is made with removable connections, such as slip-nuts and washers, the connection must be accessible. This normally is not a problem for sinks and lavatories, but it can create some problems with bathtubs. Many builders and home buyers despise having an ugly access panel in the wall where their tub waste opening is located. To eliminate the need for this type of access, the tub waste opening can be connected with permanent joints, which can mean soldering a brass tub waste or gluing a plastic one. But if the tub waste opening is connected with slip-nuts, an access panel is required.

Washing machines generally receive their incoming water from boiler drains or laundry faucets. There is a high risk of a cross-connection when these devices are used with an automatic clothes washer. This type of connection must be protected against back-siphonage. The drainage from a washing machine must be handled by an indirect waste opening. An air-break is required and is usually accomplished by placing the washer's discharge hose into a 2-in. pipe as an indirect waste receptor. The water supply to a bidet must also be protected against back-siphonage.

Dishwashers are another likely source of back-siphonage. These appliances must be equipped by installing either a back-flow protector or an air gap on the water supply piping. The drainage from dishwashers is handled differently in each zone.

Zone one requires the use of an air gap on the drainage of a dishwasher. The air gaps are normally mounted on the countertop or in the rim of the kitchen sink. The air gap forces the waste discharge of the dishwasher through open air and down a separate discharge hose. This eliminates the possibility of back-siphonage or a backup from the drainage system into the dishwasher.

Zone two requires dishwasher drainage to be separately trapped and vented or to be discharged indirectly into a properly trapped and vented fixture.

Zone three allows the discharge hose from a dishwasher to enter the drainage system in several ways. It may be individually trapped. It may discharge into a trapped fixture. The discharge hose can be connected to a wye tailpiece in the kitchen sink drainage, or it may be connected to the waste connection provided on many garbage disposers.

While we are on the subject of garbage disposers, be advised that garbage disposers require a drain of at least 1½ in. and must be trapped. It may seem to go without saying, but garbage disposers must have a water source. This doesn't mean you have to pipe a water supply to the disposer; a kitchen faucet provides adequate water supply to satisfy the code.

Floor drains must have a minimum diameter of 2 in. Remember, piping run under a floor may never be smaller than 2 in. in diameter. Floor drains must be trapped, usually must be vented, and must be equipped with removable strainers. It is necessary to install floor drains so that the removable strainer is readily accessible.

Laundry trays are required to have 1½ in. drains that should be equipped with cross-bars or a strainer. Laundry trays may act as indirect waste receptors for clothes washers. In the case of a multiple-bowl laundry tray, the use of a continuous waste is acceptable.

Lavatories are required to have drains of at least 1¼ in. diameter. The drain must be equipped with some device to prevent foreign objects from entering it. These devices can include pop-up assemblies, cross-bars, and strainers.

When installing a shower, it is necessary to secure the pipe serving the shower head with water. This riser is normally secured with a drop-ear ell and screws. It is, however, acceptable to secure the pipe with a pipe clamp.

When we talk of showers here, we are speaking only of showers, not tub-shower combinations. The frequent use of tub-shower combinations confuses many people. A shower has different requirements from those of a tub-shower combination. A shower drain must have a diameter of at least 2 in. The reason for this is simple. In a tub-shower combination, a 1½-in. drain is sufficient because the walls of the bathtub retain water until a smaller drain can remove it. A shower doesn't have high retaining walls; therefore, a larger drain is needed to clear the shower base of water more quickly. Shower drains must have removable strainers, which should have a diameter of at least 3 in.

In zone three, all showers must contain a minimum of 900 in.^{2} of shower base. This area must not be less than 30 in. in any direction. These measurements must be taken at the top of the threshold, and they must be interior measurements. A shower advertised as a 30-in. shower may not meet code requirements. If the measurements are taken from the outside dimensions, the stall will not pass muster. There is one exception to this rule. Square showers with a rough-in of 32 in. may be allowed, but the exterior of the base may not measure less than 31½ in.

Zone one requires the minimum interior area of a shower base to be at least 1024 in.^{2} When determining the size of the shower base, take the measurements from a height equal to the top of the threshold. The minimum size requirements must be maintained for a vertical height equal to 70 in. above the drain. The only objects allowed to protrude into this space are grab bars, faucets, and shower heads.

The waterproof wall enclosure of a shower or of a tub-shower combination must extend from the finished floor to a height of no less than 6 ft, or it must extend at least 70 in. above the height of the drain opening. The enclosure walls must be at the higher of the two determining factors. An example is a deck-mounted bathing unit. With a tub mounted in an elevated platform, an enclosure that extends 6 ft above the finished floor might not meet the criterion of being 70 in. above the drain opening.

Though not as common as they once were, built-up shower stalls are still popular in high-end housing. These stalls typically use a concrete base covered with tile. You may never install one of these classic shower bases, but you need to know how, just in case the need arises. These bases are often referred to as *shower pans*. Cement is poured into the pan to create a base for ceramic tile. Before you can form these pans, you must pay attention to the surface that is to be under the pan. The subfloor or other supporting surface must be smooth and able to accommodate the weight of the shower. When the substructure is satisfactory, you are ready to make your shower pan.

Shower pans must be made from a waterproof material. In the old days, these pans were made of lead or copper. Today, they are generally made with coated papers or vinyl materials. These flexible materials make the job much easier. When forming a shower pan, extend the edges of the pan material at least 2 in. above the height of the threshold. Zone one requires the material to extend at least 3 in. above the threshold. The pan material must also be securely attached to the stud walls.

Zone one is more stringent in its shower regulations. In zone one, the shower threshold must be 1 in. lower than the other sides of the shower base, but the threshold must never be lower than 2 in. The threshold must also never be higher than 9 in. When a shower is installed for handicap facilities, the threshold must be eliminated.

Zone one goes on to require the shower base to slope toward the drain with a minimum pitch of 1/4 in./ft but not more than 1/2 in./ft. The opening into the shower must be large enough to accept a shower door with a minimum width of 22 in.

The drains for this type of shower base are new to many young plumbers. Plumbers who have worked under my supervision have attempted to use standard shower drains for these types of bases. You cannot do that—at least, not if you don't want the pan to leak. This type of shower base requires a drain that is similar to some floor drains. The drain must be installed in a way that will not allow water that might collect in the pan to seep around the drain and down the exterior of the pipe. Any water entering the pan must go down the drain. The proper drain has a flange that sits beneath the pan material. The pan material must be cut to allow water into the drain. Then another part of the drain is placed over the pan material and bolted to the bottom flange. The compression of the top piece and the bottom flange, with the pan material wedged between them, creates a watertight seal. Then the strainer portion of the drain is screwed into the bottom flange housing. Since the strainer is on a threaded extension, it can be screwed up or down to accommodate the level of the finished shower pan.

Sinks are required to have drains with a minimum diameter of 1½ in. Strainers or cross-bars are required in the sink drain. If you look, you will see that basket strainers have the basket part as a strainer and also cross-bars below the basket. This provides protection from foreign objects, even when the basket is removed. If a sink is equipped with a garbage disposer, the drain opening in the sink should have a diameter of at least 3½ in.

Toilets installed in zone three are required to be water-saver models. The older models, which use 5 gal per flush, are no longer allowed in zone three for new installations.

The seat on a residential toilet must be smooth and sized for the type of toilet it serves. This usually means that the seat has a round front.

The fill valve or ballcock for toilets must be of the anti-siphon variety. Toilets of the flush-tank type are required to be equipped with overflow tubes that do double duty as re-fill conduits. The overflow tube must be large enough to accommodate the maximum water intake entering the toilet at any given time.

Whirlpool tubs must be installed as recommended by the manufacturer. All whirlpool tubs must be installed to allow access to the unit's pump. The pump's drain should be pitched to allow the pump to empty its volume of water when the whirlpool is drained. The whirlpool pump should be positioned above the fixture's trap.

All plumbing faucets and valves using both hot and cold water must be piped in a uniform manner: hot water to be piped to the left side of the faucet or valve and cold water to the right side of the faucet or valve. This uniformity reduces the risk of burns from hot water.

In zone three, valves or faucets used for showers must be designed to provide protection from scalding. Any valve or faucet used in a shower must be pressure-balanced or contain a thermostatic mixing valve. The temperature control must not allow the water temperature to exceed 110°F. This provides safety, especially to the elderly and the very young, against scalding injuries from the shower. Zones one and two do not require temperature-controlled valves in residential dwellings. When zone one requires temperature-controlled shower valves, the maximum allowable temperature is 120°F.

COMMERCIAL FIXTURE APPLICATIONS

Drinking fountains are a common fixture in commercial applications. Restaurants use garbages disposers that are so big it can take two plumbers to move them. Gang showers are not uncommon in school gyms and health clubs. Urinals are another common commercial fixture. Then there are toilets, which in commercial applications often differ from residential toilets. Special fixtures and applications exist for some unusual plumbing fixtures, such as baptismal pools in churches. This section of the chapter takes you into the commercial field and shows you how plumbing needs vary from residential uses to commercial applications.

Let's start with drinking fountains and watercoolers. The main fact to remember about watercoolers and fountains is this: They are not allowed in toilet facilities. You may not install a water fountain in a room that contains a toilet. If the building such as a restaurant, for which a plumbing diagram is being designed, will serve water or if the building will provide access to bottled water, drinking fountains and watercoolers may not be required.

As said earlier, commercial garbage disposers can be big. These monster grinding machines require a drain with a diameter of no less than 2 in. Commercial disposers must have their own drainage piping and trap. As with residential disposers, commercial disposers must have a cold-water source. In zone two, the water source must be of an automatic type. These large disposers must not be connected to a grease interceptor.

Garbage-can washers are not something you will find in the average house, but they are not uncommon to commercial applications. Owing to the nature of this fixture, its water supply must be protected against back-siphonage, which can be achieved by using either a back-flow preventer or an air gap. The waste pipes from these fixtures must have individual traps. The receptor that collects the residue from the garbage-can washer must be equipped with a removable strainer that is capable of preventing the entrance of large particles into the sanitary drainage system.

Special fixtures include church baptismal pools, swimming pools, fish ponds, and other such arrangements. The water pipes to any of these special fixtures must be protected against back-siphonage.

Showers for commercial or public use can be very different from those found in a residence. It is not unusual for showers in commercial-grade plumbing to be gang showers, which are one large shower enclosure with many shower heads and shower valves. In gang showers, the shower floor must be properly graded toward the shower drain or drains. The floor must be graded in a way to prevent water generated at one shower station from passing through the floor area of another shower station. The methods employed to divert water from each shower station to a drain are up to the designer, but it is imperative that water used by one occupant not pass into another bather's space. Zone one requires the gutters of gang showers to have rounded corners, and they must have a minimum slope toward the drain of 2 percent. The drains in the gutter must not be more than 8 ft from sidewalls and not more than 16 ft apart.

Urinals are not a common household item, but they are typical fixtures in public toilet facilities. The amount of water used by a urinal in a single flush should be limited to a maximum of 1½ gal. Water supplies to urinals must be protected from back-flow. Only one urinal may be flushed by a single flush valve. When urinals are used, they must not take the place of more than half of the toilets normally required. Urinals for public use are required to have a water trap seal that is visible and unobstructed by strainers.

Floor and wall conditions around urinals are another factor to be considered. These areas are required to be waterproof and smooth. They must be easy to clean, and they must not be made from an absorbent material. In zone three, these materials are required around a urinal in several directions. They must extend to at least 1 ft on each side of the urinal. This measurement is taken from the outside edge of the fixture. The material is required to extend from the finished floor to a point 4 ft off the finished floor. The floor under a urinal must be made of this same type of material, and the material must extend to a point at least 1 ft in front of the farthest portion of the urinal.

Commercial-grade toilets can present some of their own variations on residential requirements. The toilets used in public facilities must have elongated bowls, which must be equipped with elongated seats. Furthermore, the seats must be hinged, and they must have open or split fronts.

Flush valves are used almost exclusively with commercial-grade fixtures—toilets, urinals, and some special sinks. If a fixture depends on trap siphonage to empty itself, it must be equipped with a flush valve or a properly rated flush tank. Such valves or tanks are required for each fixture in use.

Flush valves must be equipped with vacuum breakers that are accessible. Flush valves in zone three must be rated as water-conserving valves. They must be able to be regulated for water pressure, and they must open and close fully. If water pressure is not sufficient to operate a flush valve, other measures, such as a flush tank, must be incorporated into the design. All manually operated flush tanks should be controlled by an automatic filler designed to refill the flush tank after each use. The automatic filler must be equipped to cut itself off when the trap seal is replenished and the flush tank is full. If a flush tank is designed to flush automatically, the filler device is controlled by a timer.

FIXTURES FOR HEALTH CARE

There is an entire group of special fixtures that are normally found only in facilities providing health care. The requirements for these fixtures are extensive. Although you may never have a need to work with these specialized fixtures, you should know the code requirements for them. This section of the chapter provides you with the information you may need.

Many special fixtures are required to be made of materials providing a higher standard than materials for normal fixtures. They may be required to endure excessive heat or cold. Many of them are also required to be protected against back-flow. The prohibition of back-flow extends to the drainage system, as well as to the potable water supply. All special fixtures must be of an approved type.

Sterilizers

Any concealed piping that serves special fixtures and that may require maintenance or inspection must be accessible. All piping for sterilizers must be accessible. Steam piping to a sterilizer should be installed with a gravity system to control condensation and to prevent moisture from entering the sterilizer. Sterilizers must be equipped with a means to control the steam vapors. The drains from sterilizers must be piped as indirect wastes. Sterilizers are required to have leak detectors designed to expose leaks and to carry unsterile water away from the sterilizer. The interior of sterilizers may not be cleaned with acid or other chemical solutions while the sterilizers are connected to the plumbing system.

Clinical Sinks

Clinical sinks are sometimes called *bedpan washers*. Clinical sinks are required to have an integral trap. The trap seal must be visible, and the contents of the sink must be removed by siphonic or blowout action. The trap seal must be automatically replenished, and the sides of the fixture must be cleaned by a flush rim at every flushing of the sink. These special fixtures are required to connect to the DWV system in the same manner as a water closet. When clinical sinks are installed in utility rooms, they are not meant to be a substitute for a service sink. On the other hand, service sinks may never be used to replace a clinical sink. Devices for making or storing ice are not allowed in a soiled utility room.

Vacuum Fluid-Suction Systems

Vacuum system receptacles are to be built into cabinets or cavities, but they must be visible and readily accessible. Bottle suction systems used for collecting blood and other human fluids must be equipped with overflow prevention devices at each vacuum receptacle. Secondary safety receptacles are recommended as an additional safeguard. Central fluid-suction systems must provide continuous service. If a central suction system requires periodic cleaning or maintenance, it must be installed so that it can continue to operate even while cleaning or maintenance is being performed. When central systems are installed in hospitals, they must be connected to emergency power facilities. The vent discharge from these systems must be piped separately to the outside air above the roof of the building.

Waste originating in a fluid-suction system that is to be drained into the normal drainage piping must be piped into the drainage system with a direct-connect, trapped arrangement; indirect waste connection of this type of unit is not allowed. Piping for fluid-suction systems must be noncorrosive and have a smooth interior surface. The main pipe must have a diameter of no less than 1 in. Branch pipes must not be smaller than 1/2 in. All piping is required to have accessible clean-outs and must be sized according to manufacturers' recommendations. The air flow in a central fluid-suction system should not be allowed to exceed 5000 ft/min.

Special Vents

Institutional plumbing uses different styles of vents for some equipment from those encountered with normal plumbing. One such vent is called a *local vent*. One example of use for a local vent is bedpan washers (clinical sinks). A bedpan washer must be connected to at least one vent, with a minimum diameter of 2 in., and that vent must extend to the outside air above the roof of the building.

Local vents are used to vent odors and vapors. Local vents may not tie in with vents from the sanitary plumbing or sterilizer vents. In multistory buildings, a local vent stack may be used to collect the discharge from individual local vents for multiple bedpan washers located above each other. A 2-in. stack can accept up to three bedpan washers. A 3-in. stack can handle six units, and a 4-in. stack can accommodate up to 12 bedpan washers. Local vent stacks are meant to tie into the sanitary drainage system, and they must be vented and trapped if they serve more than one fixture.

Each local vent must receive water to maintain its trap seal. The water source comes from the water supply for the bedpan washer being served by the local vent. A minimum of a 1/4-in. tubing must be run to the local vent, and it must discharge water into the vent each time the bedpan washer is flushed.

Vents serving multiple sterilizers must be connected with inverted wye fittings, and all connections must be accessible. Sterilizer vents are intended to drain to an indirect waste. The minimum diameter of a vent for a bedpan sterilizer must be 1½ in. When serving a utensil sterilizer, the minimum vent size must be 2 in. Vents for pressure-type sterilizers must be at least 2½ in. in diameter. When serving a pressure instrument sterilizer, a vent stack must be at least 2 in. in diameter. Up to two sterilizers of this type may be on a 2-in. vent. A 3-in. stack can handle four units.

Dual Water Service

Hospitals are required to have at least two water services. The two water services may, however, connect to a single water main. Hot water must be made available to all fixtures, as required by the fixture manufacturer. All water heaters and storage tanks must be of a type approved for the intended use.

Zone two requires the hot water system to be capable of delivering 6½ gal of 125°F water per hour for each bed in a hospital. Zone two further requires hospital kitchens to have a hot-water supply of 180°F water equal to 4 gal/hr for each bed. Laundry rooms are required to have a supply of

180°F water at a rate of 4½ gal/hr for each bed. Zone two continues its hot-water regulations by requiring hot-water storage tanks to have capacities equal to no less than 8 percent of the water-heating capacity.

Zone two continues its hot-water requirements by dictating the use of copper in submerged steam-heating coils. If a building is higher than three levels, the hot-water system must be equipped to circulate. Valves are required on the water distribution piping to fixture groups.

Back-Flow Prevention

When back-flow prevention devices are installed, they must be at least 6 in. above the flood-level rim of the fixture. In the case of hand-held showers, the height of installation must be 6 in. above the highest point at which the hose can be used.

In most cases, hospital fixtures are protected against back-flow by the use of vacuum breakers. However, a boiling type of sterilizer should be protected with an air gap. Vacuum suction systems may be protected by either an air gap or a vacuum breaker.

12

Water Heaters

Most plumbing systems involve the use of water heaters. There is really nothing very complex about selecting or installing a water heater, but some plumbers have trouble with these fixtures. Most of their confusion relates to specific code issues such as sizing a temperature and pressure relief valve or installing a safety pan. Read this chapter to get yourself up to speed on water heaters.

WORKING PRESSURE

The standard working pressure for a water heater is 125 psi. The maximum working pressure of a water heater is required to be permanently marked in an accessible location. Every water heater is required to have a drain, which is located at the lowest possible point on the water heater. Some exceptions may be allowed for very small, under-the-counter water heaters.

All water heaters are required to be insulated. The insulation factors are determined by the heat loss of the tank in 1 hr. These requirements must be met before a water heater is approved for installation.

Relief valves are mandatory equipment on water heaters. Safety valves are designed to protect against excessive temperature and pressure. The most common type of safety valve used protects against both temperature and pressure from a single valve. The blowoff rating for these valves must not exceed 210°F and 150 psi. As for the pressure relief valve rating, the valve must not have a blowoff rating of more than the maximum working pressure of the water heater it serves—usually 125 psi.

When temperature and pressure relief valves are installed, their sensors should monitor the top 6 in. of water in the water heater. Valves located between the water heater and the temperature and pressure relief valves are not allowed.

To protect bystanders in the event of a blowoff, the blowoff from relief valves must be piped down (Fig. 12.1). The pipe used for this purpose must be rigid and capable of sustaining temperatures of up to 210°F. The discharge pipe must be the same size as the relief valve's discharge opening, and it must run, undiminished in size, to within 6 in. of the floor. If a relief valve discharge pipe is piped into the sanitary drainage system, the connection must be through an indirect waste. The end of a discharge pipe may not be threaded, and no valves may be installed in the discharge pipe.

FIGURE 12.1   SCHEMATIC DESIGN OF WATER HEATER

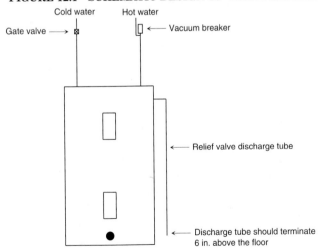

When the discharge from a relief valve might damage property or people, safety pans should be installed. These pans typically have a minimum depth of 1½ in. Plastic pans are commonly used for electric water heaters, and metal pans are used for fuel-burning heaters. The pans must be large enough to accommodate the discharge flow from the relief valve.

The pan's drain may be piped to the outside of the building or to an indirect waste where people and property will not be affected. The discharge location should be chosen so that it will be obvious to building occupants when a relief valve discharges. Traps should not be installed on the discharge piping from safety pans.

Water heaters must be equipped with an adjustable temperature control, which is required to be automatically adjustable from the lowest to the highest temperatures allowed. Some locations restrict the maximum water temperature in residences to 120°F. A switch must be supplied to shut off the power to electric water heaters. When the water heater uses a fuel, such as gas, a valve must be available to cut off the fuel source. Both the electric and fuel shutoffs must be able to be used without affecting the remainder of the building's power or fuel. All water heaters requiring venting must be vented in compliance with local code requirements.

TROUBLESHOOTING WATER HEATERS IN GENERAL

Problem	Probable Cause	Remedy
Relief valve drips	Excessive temperature setting	Lower setting
	Defective valve	Replace
High operating costs	Sediment, rust, or lime in tank	Drain and flush (replace heater if buildup is severe)
	Water heater too small for job	Replace with larger heater
	Wrong size piping connections	Install correct size piping

TROUBLESHOOTING WATER HEATERS IN GENERAL *(cont.)*

Problem	Probable Cause	Remedy
	Lack of insulation on long runs of pipe	Insulate pipes
Condensation	Heater installed in closed or confined area	Vent room or use louvers to allow air circulation
Water tank leaks	Rusting of inner tank walls causing pin holes	Replace heater (any repair is only temporary)
Rusty hot water	Buildup of rust in heater	Drain and flush; if problem is severe, replace heater

TROUBLESHOOTING ELECTRIC WATER HEATERS

Problem	Probable Cause	Remedy
No hot water	Blown fuses or defective circuit breaker	Replace or adjust
	Defective heating element	Replace
	Broken thermostat	Replace
	Broken time clock	Replace
Hot water turns cold quickly	Defective upper or lower heating elements	Replace defective parts
	Defective thermostat	Replace
	Rust buildup	Drain and clean heater
Extremely hot water comes out of faucet	Defective thermostat	Replace

TROUBLESHOOTING GAS WATER HEATERS

Problem	Probable Cause	Remedy
Gas smell	Hole in fittings, valves, flue, or connections	Replace any worn or defective parts; tighten connections
No gas	No gas coming in	Check valves and adjust or replace
	Defective thermo-coupling	Replace
	Dip tube installed incorrectly	Adjust—tube must be in the cold water supply
Extremely hot water	Defective gas control valve	Replace
Water turns cold quickly	Dip tube in wrong inlet or broken	Insert tube in cold water supply
	Gas controller defective	Replace defective parts

13

Back-Flow Prevention

Back-flow prevention has always been a serious issue, but it has come to be a primary concern in modern plumbing systems—with good reason. A plumbing system that is not protected adequately from back-flow can result in disaster. People can become ill or even fatally injured when back-flow occurs. The plumbing code is very strict on this issue, so you had better know the rules and regulations. This chapter apprises you of the key concerns.

An air gap is the most positive form of protection from back-flow. However, air gaps are not always feasible. Because air gaps cannot always be used, a number of devices are available for the protection of potable water systems.

VENTED PREVENTERS

Some back-flow preventers are equipped with vents. When these devices are used, the vents must not be installed so that they may become submerged. It is also required that these units be capable of performing their function under continuous pressure. Other back-flow preventers are designed to operate in a manner similar to that of an air gap. When conditions occur that may cause a back-flow, these devices open and create an open air space between the two pipes connected to them. Reduced-pressure back-flow preventers perform this action very well.

Another type of back-flow preventer that performs in a similar way is an atmospheric vent back-flow preventer. When potable water is connected to a boiler for heating purposes, the potable water inlet should be equipped with a vented back-flow preventer. If the boiler contains chemicals in its water, the potable water connection should be made with an air gap or a back-flow preventer that operates on the principle of reduced pressure.

VACUUM BREAKERS

Vacuum breakers are frequently installed on water heaters, hose bibbs, and sillcocks to prevent back-flow. They are also installed on the faucet spout of laundry tubs. These devices either mount on a pipe or screw onto a hose connection. Some sillcocks are equipped with factory-installed vacuum breakers. These devices open when necessary and break any siphonic action with the introduction of air.

When vacuum breakers are installed, they must be installed at least 6 in. above the flood-level rim of the fixture. Because of the way they are designed to introduce air into the potable water piping, vacuum breakers may not be installed where they may suck in toxic vapors or fumes. For example, it would not be acceptable to install a vacuum breaker under the exhaust hood of a kitchen range.

Lawn sprinklers and irrigation systems must be installed with back-flow prevention in mind. Vacuum breakers are a preferred method for back-flow prevention, but other types of back-flow preventers are allowed in such systems.

A BAROMETRIC LOOP

In some specialized cases a barometric loop is used to prevent back-siphonage. In zone three, the loop must extend at least 35 ft high and can be used only as a vacuum breaker. These loops are effective because they rise higher than the point at which a vacuum suction can occur. Barometric loops work because by being 35 ft in height, suction cannot be achieved.

DOUBLE CHECK VALVES

Double check valves are used in some instances to control back-flow. When used in this capacity, double check valves must be equipped with approved vents. This type of protection would be used on a carbonated beverage dispenser, for example.

Connections between a potable water supply and an automatic fire sprinkling system should be made with a check valve. If the potable water supply is being connected to a nonpotable water source, the connection should be made through a back-flow preventer that works on the principle of reduced pressure.

AIR GAP

Some fixtures require an air gap as protection from back-flow. Examples of these fixtures are the following:

Lavatories   Sinks

Laundry tubs   Bathtubs

Drinking fountains

The air gap for the these fixtures is accomplished through the design and installation of the faucet or spout serving them.

INSPECTION

Back-flow preventers require inspection from time to time. Therefore, they must be installed in accessible locations.

INDIRECT WASTE

Indirect waste requirements can pertain to a number of types of plumbing fixtures and equipment. You might not think of a standpipe for a clothes-washing machine as a back-flow preventer, but in a way it is. Indirect waste connections in drainage systems prevent contamination of plumbing fixtures. These wastes can include a clothes-washer drain, a condensate line, a sink drain, or the blowoff pipe from a relief valve, just to name a few. Indirect wastes are piped in this manner to prevent, among other things, the possibility of contaminated matter backing up the drain into a potable water or food source.

Most indirect waste receptors are trapped. If the drain from the fixture is more than 2 ft long, the indirect waste receptor must be trapped. However, this trap rule applies to fixtures like sinks, not to an item such as a blowoff pipe from a relief valve. The rule is different in zone one. In zone one, if the drain is more than 5 ft long, it must be trapped.

The safest method of creating an indirect waste preventer is accomplished by using an air gap. When an air gap is used, the drain from the fixture terminates above the indirect waste receptor, with open-air space between the waste receptor and the drain. This prevents any backup or back-siphonage.

Some fixtures, depending on local code requirements, may be piped with an air break rather than an air gap. With an air break, the drain may extend below the flood-level rim and terminate just above the trap's seal. The risk in using an air break is the possibility of a backup. Since the drain is run below the flood-level rim of the waste receptor, it is possible that the waste receptor could overflow and back up into the drain. This could create contamination, but in cases where contamination is likely, an air gap will be required. Check with your local codes office before using an air break.

STANDPIPES

Standpipes, such as those used for washing machines, are a form of indirect waste receptors. A standpipe used for this purpose in zones one and three must extend at least 18 in. above the trap's seal, but it may not extend more than 30 in. above the seal. If a clear-water waste receptor is located in a floor, zone three requires the lip of the receptor to extend at least 2 in. above the floor to eliminate the waste receptor from being used as a floor drain.

Choosing the proper size for a waste receptor is generally based on the receptor's ability to handle the discharge from a drain without excessive splashing. If you are concerned with sizing a particular waste receptor, talk with your local code officer for a ruling.

Food Preparation

Buildings used for food preparation, food storage, and similar activities are required to have their fixtures and equipment discharge drainage through an air gap. Zone three provides an exception to this rule. In zone three, dishwashers and open culinary sinks are excepted. Zone two requires the discharge pipe to terminate at least 2 in. above the receptor. Zone one requires the distance to be a minimum of 1 in. Zones two and three require the air gap distance to be a minimum of twice the size of the pipe discharging the waste. For example, a 1/2-in. discharge pipe would require a 1-in. air gap.

Zones two and three prohibit the installation of an indirect waste receptor in any room containing toilet facilities. Zone one goes along with this ruling but allows one exception. The exception is the installation of a receptor for a clothes washer when the clothes washer is installed in the same room. Indirect waste receptors are not allowed to be installed in closets and other unvented areas. Indirect waste receptors must be accessible. Zone two requires all receptors to be equipped with a means of preventing solids with diameters of 1/2 in. or larger from entering the drainage system. These straining devices must be removable to allow for cleaning.

Extreme Water Temperatures

When you deal with extreme water temperatures in waste water, such as with a commercial dishwasher, the dishwasher drain must be piped to an indirect waste. The indirect waste will be connected to the sanitary plumbing system, but the dishwasher drain must not connect to the sanitary system directly if the waste-water temperature exceeds 140°F. Steam pipes must not be connected directly to a sanitary drainage system. Local regulations may require the use of special piping, sumps, or condensers to accept high-temperature water. Zone one prohibits the direct connection of any dishwasher to the sanitary drainage system.

Clear Water

Clear water waste from a potable source must be piped to an indirect waste with the use of an air gap. Sterilizers and swimming pools are two examples of when this rule would apply. Clear water from nonpotable sources, such as a drip from a piece of equipment, must be piped to an indirect waste. In zone three, an air break is allowed in place of an air gap. Zone two requires any waste entering the sanitary drainage system from an airconditioner to do so through an indirect waste.

14

Drainage Tips

This chapter provides you with tips on drainage systems. Although it is not a comprehensive listing of all code requirements pertaining to drainage systems, it contains many useful tidbits. I've selected some of the elements of drainage systems that plumbers tend to ask the most questions about. The information should prove helpful to you when a quick answer to a common question is needed.

PIPE SUPPORTS

Pipe supports are regulated by the plumbing code. One concern with the type of hangers used is their compatibility with the pipe they are supporting. You must use a hanger that will not have a detrimental effect on your piping. For example, you may not use a galvanized straphanger to support copper pipe. As a rule of thumb, the hangers used to support a pipe should be made from the same material as the pipe being supported. For example, copper pipe should be hung with copper hangers. This eliminates the risk of a corrosive action between two different types of materials. You may use a plastic or plastic-coated hangers with all types of pipe. The exception to this rule is when the piping is carrying a liquid with a temperature that might affect or melt the plastic hanger.

The hangers used to support pipe must be capable of supporting the pipe at all times. Hangers must be attached to pipes and to the members holding the hangers in a satisfactory manner. For example, it would not be acceptable to wrap a piece of wire around a pipe and then wrap the wire around the bridging between two floor joists. Hangers should be securely attached to the members supporting them. For example, a hanger should be attached to the pipe and then nailed to a floor joist. The nails used to hold a hanger in place should be made of the same material as the hanger if corrosive action is a possibility.

Both horizontal and vertical pipes require support. The intervals between supports varies, depending upon the type of pipe being used and whether it is installed vertically or horizontally. The following examples show the frequency with which the various types of pipes must be supported when they are hung horizontally. The following examples are the maximum distances allowed between supports for zone three:

ABS—4 ft   Cast iron—5 ft

Galvanized steel—12 ft   PVC—4 ft

DWV copper—10 ft

When these same types of pipes are installed vertically in zone three, they must be supported at no less than the following intervals:

ABS—4 ft   Cast iron—15 ft

Galvanized steel—15 ft   PVC—4 ft

DWV copper—10 ft

When installing cast-iron stacks, you must support the base of each stack because of the weight of cast-iron pipe. If you are installing pipe with flexible couplings, bands, or unions, you must install and support the pipe in a way that prevents these flexible connections from moving. In pipes larger in diameter than 4 in., all flexible couplings must be supported to prevent the force of the pipe's flow from loosening the connections at changes in direction.

CLEAN-OUTS

Many places in a plumbing system require clean-outs. Let's start with sewers. All sewers must have clean-outs. The distances between these clean-outs vary from region to region. Generally, clean-outs are required where the building drain meets the building sewer. The clean-outs may be installed inside the foundation or outside, but its opening must extend upward to the finished floor level or the finished grade outside.

Zone two requires that the clean-outs at the junction of building drains and sewers be located outside. If the clean-out is installed inside, it must extend above the flood level rim of the fixtures served by the horizontal drain. When this is not feasible, allowances may be made. Zone three will waive the requirement for a junction clean-out if there is a clean-out of at least 3-in. diameter within 10 ft of the junction.

Once the sewer is begun, clean-outs should be installed every 100 ft. In zone two, the interval distance is 75 ft for 4-in. and larger pipe and 50 ft for pipe smaller than 4 in. Clean-outs are also required in sewers when the pipe changes direction. In zone three, a clean-out is required every time the sewer turns more than 45°. In zone one, a clean-out is required whenever the change in direction is more than 135°.

The clean-outs installed in a sewer must be accessible, which generally means that a standpipe will rise from the sewer to just below ground level. At that point, a clean-out fitting and plug are installed on the standpipe. This allows the sewer to be snaked out from ground level, with little to no digging required.

Clean-out Plugs

Clean-out plugs are made of plastic, brass, or borosilicate-glass. Brass plugs are used only with metallic fittings. Unless they create a hazard, clean-out plugs have raised, square heads. If located where a hazard from the raised head may exist, countersunk heads may be used. Zone two requires borosilicate-glass plugs to be used with clean-outs installed on borosilicate pipe.

Building Drains

For building drains and horizontal branches, clean-out location depends upon pipe size, but clean-outs are normally required every 50 ft. For pipes with diameters of 4 in. or less, clean-outs must be installed every 50 ft. Larger drains may have clean-outs spaced at 100-ft intervals. Clean-outs are also required on these pipes with a change in direction. For zone three, the degree of change is anything in excess of 45°. Clean-outs must be installed at the end of all horizontal drain runs. Zone one does not require clean-outs at 50-ft intervals—only 100-ft intervals.

As with most rules, there are some exceptions. Zone one offers some exceptions to the clean-out requirements for horizontal drains. The following exceptions apply only to zone one. If a drain is less than 5 ft long and does not drain sinks or urinals, a clean-out is not required. A change in direction from a vertical drain with a sixth bend does not require a clean-out. Clean-outs are not required on pipes other than building drains and their horizontal branches that are above the first-floor level.

P-Traps

P-traps and water closets are often allowed to act as clean-outs. When these devices are approved for clean-out purposes, the normally required clean-out fitting and plug at the end of a horizontal pipe run may be eliminated. Not all jurisdictions accept P-traps and toilets as clean-outs; check your local requirements before omitting standard clean-outs.

A clean-out must be installed in a way that makes its opening accessible and allows adequate room for drain cleaning. The clean-out must be installed to go with the flow. This means that when the clean-out plug is removed, a drain-cleaning device should be able to enter the fitting and the flow of the drainage pipe without trouble.

When you install plumbing in zone three, you must include a clean-out at the base of every stack. This is good procedure at any time, but it is not required by all codes. The height of this clean-out should not exceed 4 ft. Many plumbers install test tees at these locations to plug their stacks for pressure testing. The test tee doubles as a clean-out.

When the pipes holding clean-outs is concealed, the clean-out must be made accessible. For example, if a stack is to be concealed by a finished wall, provision must be made for access to the clean-out. Access can take the form of an access door, or the clean-out can simply extend past the finished wall covering. If the clean-out serves a pipe concealed by a floor, the clean-out must be brought up to floor level and made accessible. This ruling applies not only to clean-outs installed beneath concrete floors but also to clean-outs installed in crawlspaces where there is very little room to work.

Size

Size is one of the lessons to be learned about clean-outs. All clean-outs are required to be the same size as the pipe they are serving unless the pipe is larger than 4 in. in diameter. If you install a 2-in. pipe, you must install 2-in. clean-outs. However, when a P-trap is allowed for a clean-out, it may be smaller than the drain. An example would be a 1¼-in. trap on an 1½-in. drain. Remember, though, that not all code enforcement officers will allow P-traps as clean-outs, and they may require the P-trap to be the same size as the drain if the trap is allowed as a clean-out. Once the pipe size exceeds 4 in., the clean-outs used should have a minimum size of 4 in.

Adequate Clearance

When clean-outs are installed, there must be adequate clearance in front of them for drain cleaning. The clearance required for pipes with diameters of 3 in. or more is 18 in. Smaller pipes require a minimum clearance of 12 in. Many plumbers fail to remember this regulation. It is common to find clean-outs pointing toward floor joists or too close to walls. You will save yourself time and money by committing these clearance distances to memory.

Zone one takes the clearance rules a step further. When a clean-out is installed in a floor, it must have a minimum vertical clearance of 18 in. and a minimum horizontal clearance of 30 in. No under-floor clean-out is allowed to be placed more than 20 ft from an access opening.

Clean-out plugs and plates must be easily removed. Access to the interior of the pipe should be available without undue effort or time. Clean-outs can take on many appearances. The U bend of a P-trap can be considered a clean-out, depending upon local interpretation. A rubber cap held onto the pipe by a stainless steel clamp can serve as a clean-out. A standard female adapter and plug make a fine clean-out. Test tees will work as clean-outs. Special clean-outs designed to allow rodding of a drain in either direction are acceptable.

Manholes

The largest clean-out is a manhole. When a pipe's diameter exceeds 10 in. in zone three or 8 in. in zone two, manholes replace clean-outs. Manholes are required every 400 ft in zone three and every 300 ft in zones one and two. In addition, they are required at all changes in direction, elevation, grade, and size. Manholes must be protected against flooding and must be equipped with covers to prevent the escape of gases. Zone one requires connections with manholes to be made with flexible compression joints. Connections must not be made closer than 1 ft to the manhole and not farther than 3 ft away from it.

OFFSETS IN HORIZONTAL PIPING

When you want to change the direction of a horizontal pipe, you must use fittings approved for that purpose. You have six choices in zone three, as follows:

Sixteenth bend	Eighth bend
Sixth bend	Long-sweep fittings
Combination wye and eighth bend	Wye

GOING FROM HORIZONTAL TO VERTICAL

You have a wider range of choice in selecting a fitting for going from a horizontal position to a vertical position. There are nine possible candidates available for this type of change in direction when you are working in zone three, as follows:

Sixteenth bend	Eighth bend
Sixth bend	Long-sweep fittings
Combination wye and eighth bend	Wye
Quarter bend	Short-sweep fittings
Sanitary tee	

You may not use a double sanitary tee in a back-to-back situation if the fixtures being served are of a blow-out or pump type. For example, you cannot use a double sanitary tee to receive the discharge of two washing machines if the machines are positioned back to back. The sanitary tee's throat is not deep enough to keep drainage from feeding back and forth between the fittings. In a case like this, use a double combination wye and eighth bend. The combination fitting has a much longer throat and will prohibit waste water from transferring across the fitting to the other fixture.

VERTICAL TO HORIZONTAL CHANGES IN DIRECTION

Seven fittings are allowed to change direction from vertical to horizontal. They are as follows:

Sixteenth bend	Eighth bend
Sixth bend	Long-sweep fittings
Combination wye and eighth bend	Wye

Short-sweep fittings that are 3 in. or larger

Zone one prohibits a fixture outlet connection within 8 ft of a vertical-to-horizontal change in direction of a stack if the stack serves a suds-producing fixture. A suds-producing fixture could be a laundry fixture, a dishwasher, a bathing unit, or a kitchen sink. This rule does not apply to single-family houses or to stacks in buildings with fewer than three stories.

INTERCEPTORS AND SEPARATORS

Interceptors are used to keep harmful substances from entering the sanitary drainage systems. Because they separate the materials entering the system and retain certain materials, while allowing others to continue into the drainage system, separators are also required in some circumstances. Interceptors are used to control grease, sand, oil, and other materials.

Interceptors and separators are required when conditions provide opportunity for harmful or unwanted materials to enter a sanitary drainage system. For example, a restaurant is required to be equipped with a grease interceptor because of the large amount of grease present in commercial food establishments. An oil separator is required for a building in which automotive repairs are made. Interceptors and separators must be designed for each individual situation. There is no rule-of-thumb method for choosing the proper interceptor or separator without expert design.

Some guidelines are provided in plumbing codes for interceptors and separators. The capacity of a grease interceptor is based on two factors: grease retention and flow rate. These determinations are typically made by a professional designer. The size of a receptor or separator is also normally determined by a design expert.

Interceptors for sand and other heavy solids must be readily accessible for cleaning. These units must contain a water seal of not less than 6 in., except in zone two, where the minimum water depth is 2 in. When an interceptor is used in a laundry, a water seal is not required. Laundry receptors, used to catch lint, string, and other objects, are usually made of wire, and they must be easily removed for cleaning. Their purpose is to prevent solids with a diameter of 1/2 in. or more from entering the drainage system.

Other types of separators are used for various plants, factories, and processing sites. The purpose of all separators is to keep unwanted objects and substances from entering the drainage system. Vents are required if it is suspected that these devices will be subject to the loss of a trap seal. All interceptors and separators must be readily accessible for cleaning, maintenance, and repairs.

SEWER PUMPS AND PITS

There are times when conditions do not allow a plumbing system to flow by gravity in the desired direction. When this is the case, sewer pumps and pits become necessary. It is also possible that sump pumps and sumps must be used to remove water collected below the level of the building drain or sewer. When you plan to install a pump or sump pit, you must abide by certain regulations.

All sump pits must have a sealed cover that will not allow the escape of sewer gas. The pit size is determined by the size and performance of the pump to be housed in the sump. But the sump generally must have a minimum diameter of 18 in. and a minimum depth of 24 in. If a sewer pump is installed in a pit, the pump must be capable of lifting solids with a diameter of 2 in., up into a gravity drain or sewer. The discharge pipe from these sumps must have a minimum diameter of 2 in. Zone one requires the drain receiving the discharge from a sewer sump to be sized with a rating of two fixture-units for every gpm the pump is capable of producing.

If the sump pit will not receive any discharge from toilets, the pump is not required to lift the 2-in. solids and may be smaller. A standard procedure is to install a pump capable of lifting 1/2-in. solids to a gravity drain if no toilets discharge into the sump.

It is a good idea (and may be a requirement) to install two pumps in the sump. The pumps may be installed in a manner to take turns with the pumping chores, but most important, if one pump fails, the other pump can continue to operate. Zone two requires the installation of this type of two-pump system when six or more water closets discharge into the sump. Zone one requires any public installation of a sewer sump to be equipped with a two-pump system. Alarm systems are often installed on sewer pump systems to warn building occupants if the water level in the sewer pit rises to an unusually high level. Zones one and two require the efflu-ent level to remain at least 2 in. below the inlet of the sump.

All sewer sumps should be equipped with a vent. Ide-ally, the vent should extend upward to open air space with-out tying into another vent. Most sump vents are 2 in. in di-ameter, but in no case are they allowed to have a diameter of less than 1½ in.

A check valve and a gate valve should be installed on the discharge piping from the pump. These devices prevent water from running back into the sump and allow the pump to be worked on with relative ease.

Most sewer pumps are equipped with a 2-in discharge outlet. An ejector pump with a 2-in. outlet should be able to pump 21 gal/min. If the discharge outlet is 3 in. in diameter, the pump should have a flow rate of 46 gal/min.

TRAPS

Traps are required on almost all plumbing fixtures. This is a common requirement, but there are many types of traps that can be used. Let's look at some of these on an individual basis.

P-Traps

P-traps are the traps most frequently used in modern plumb-ing systems. They are self-cleaning and frequently have re-movable U-bends that may act as clean-outs pending local approval. P-traps must be properly vented. Without adequate venting, the trap seal can be removed by back-pressure.

S-Traps

S-traps were very common when most plumbing drains came up through the floor instead of out of a wall. Many S-traps are still in operation, but they are no longer allowed in new installations. S-traps are subject to losing their trap seal through self-siphoning.

Drum Traps

Drum traps are not normally allowed in new installations without special permission from the code officer. The only oc-casion when drum traps are still used frequently is when they are installed with a combination waste-and-vent system.

Bell Traps

Bell traps are not allowed in new installations.

House Traps

House traps are no longer allowed; they represent a double trapping of all fixtures. House traps were once installed where the building drain joined with the sewer. Most house traps were installed inside the structure, but a fair number were installed outside, underground. Their purpose was to prevent sewer gas from coming out of the sewer and into the plumbing system. House traps make drain cleaning very difficult, and they create a double-trapping situation, which is not allowed. This regulation, like most regulations, is subject to amendment and variance by the local code official.

Crown-Vented Traps

Crown-vented traps are not allowed in new installations. These traps have a vent rising from the top of the trap. As you learned earlier, crown venting must be done at the trap arm, not at the trap.

Other Traps

Traps that depend on moving parts or interior partitions are not allowed in new installations.

TRAP SIZES

Trap sizes are determined by the local code. A trap may not be larger than the drain pipe it discharges into.

SUPPORT INTERVALS FOR VERTICAL DRAINAGE PIPE

	Type of Pipe	*Maximum Distance Between Supports*
Zone 1	Lead	4 ft
	Cast iron	At each story
	Galvanized steel	At least every other story
	Copper	At each story[a]
	PVC	Not mentioned
	ABS	Not mentioned
Zone 2	Lead	4 ft
	Cast iron	At each story[b]
	Galvanized steel	At each story[c]
	Copper (1¼ in. and smaller)	4 ft
	Copper (1½ in. and larger)	At each story

SUPPORT INTERVALS FOR VERTICAL DRAINAGE PIPE *(cont.)*

Type of Pipe	Maximum Distance Between Supports
PVC (1½ in. and smaller)	4 ft
PVC (2 in. and larger)	At each story
ABS (1 ½ in. and smaller)	4 ft
ABS (2 in. and larger)	At each story

[a]Support intervals may not exceed 10 ft.

[b]Support intervals may not exceed 15 ft.

[c]Support intervals may not exceed 30 ft.

Note: All stacks must be supported at their bases.

SUPPORT INTERVALS FOR VERTICAL VENT PIPE

	Type of Pipe	Maximum Distance Between Supports
Zone 1	Lead	4 ft
	Cast iron	At each story
	Galvanized steel	At least every other story
	Copper	At each story[a]
	PVC	Not mentioned
	ABS	Not mentioned
Zone 2	Lead	4 ft
	Cast iron	At each story[b]
	Galvanized steel	At each story[c]
	Copper (1¼ in.)	4 ft
	Copper (1½ in. and larger)	At each story
	PVC (1½ in. and smaller)	4 ft
	PVC (2 in. and larger)	At each story
	ABS (1½ in. and smaller)	4 ft
	ABS (2 in. and larger)	At each story
Zone 3	Lead	4 ft
	Cast iron	15 ft
	Galvanized steel	15 ft
	Copper tubing	10 ft
	ABS	4 ft
	PVC	4 ft
	Brass	10 ft
	Aluminum	15 ft

[a]Support intervals may not exceed 10 ft.

[b]Support intervals may not exceed 15 ft.

[c]Support intervals may not exceed 30 ft.

Note: All stacks must be supported at their bases.

SUPPORT INTERVALS FOR HORIZONTAL VENT PIPE

	Type of Pipe	Maximum Distance Between Supports (ft)
Zone 1	ABS	4
	Cast iron	At each pipe joint[a]
	Galvanized steel	12
	Copper (1½ in. and smaller)	6
	Copper (2 in. and larger)	10
	PVC	4
Zone 2	ABS	4
	Cast iron	At each pipe joint
	Galvanized steel (1 in. and larger)	12
	PVC	4
	Copper (2 in. and larger)	10
	Copper (1½ in. and smaller)	6
Zone 3	Lead	Continuous
	Cast iron	5[a]
	Galvanized steel	12
	Copper tube (1¼ in.)	6
	Copper tube (1½ in. and larger)	10
	ABS	4
	PVC	4
	Brass	10
	Aluminum	10

[a]Cast-iron pipe must be supported at each joint, but supports may not be more than 10 ft apart.

GRADING YOUR PIPE

When you install horizontal drainage piping, you must install it so that it falls toward the waste disposal site. A typical grade for drainage pipe is 1¼ in./ft. For example, the lower end of a 20-ft piece of pipe would be 5 in. lower than the upper end when properly installed. Although the grade of a 1/4 in./ft is typical, it is not the only acceptable grade for all pipes.

If you are working with pipe that has a diameter of 2½ in. or less, the minimum grade for the pipe is 1/4 in./ft. Pipes with diameters between 3 and 6 in. are allowed a minimum grade of 1/8 in./ft. Zone one requires special permission to be granted before you can install pipe with a 1/8 in./ft grade. In zone three, an acceptable grade for pipes with diameters of 8 in. or more is 1/16 in./ft.

BACKWATER VALVES

Backwater valves are essentially check valves. They are installed in drains and sewers to prevent the backing up of waste and water in the drain or sewer. Backwater valves are required to be readily accessible and installed any time a drainage system is likely to encounter backups from the sewer.

The intent behind using backwater valves is to prevent sewers from backing up into individual drainage systems. Buildings that have plumbing fixtures below the level of the street, where a main sewer is installed, are candidates for backwater valves.

SPECIAL WASTES

Special wastes are those that may have a harmful effect on a plumbing system or the waste disposal system. Possible locations for special waste piping include photographic labs, hospitals, or buildings where chemicals or other potentially dangerous wastes are dispersed. Small, private photo darkrooms do not generally fall under the scrutiny of these regulations. Buildings that are considered to have a need for special-waste plumbing are often required to have two plumbing systems—one system for normal sanitary discharge and a separate system for the special wastes. Before many special wastes are allowed to enter a sanitary drainage system, they must be neutralized, diluted, or otherwise treated.

Depending upon the nature of the special wastes, special materials may be required. When you venture into the plumbing of special wastes, it is always best to consult the local code officer before proceeding with your work.

STANDPIPE HEIGHT

A standpipe, when installed in zone three, must extend at least 18 in. above its trap but may not extend more than 30 in. above it. Zone two prohibits the standpipe from extending more than 4 ft from the trap. Zone one requires the standpipe not to exceed a height of more than 2 ft above the trap. Plumbers installing laundry standpipes often forget this regulation. When setting your fitting height in the drainage pipe, keep in mind the height limitations on your standpipe. Otherwise, your takeoff fitting may be too low or too high to allow your standpipe receptor to be placed at the desired height. Traps for kitchen sinks may not receive the discharge from a laundry tub or clothes washer.

MINIMUM PITCH FOR DRAINAGE PIPE

	Pipe Diameter	Pitch per Foot
Zone 1	Under 4 in.	1/4 in.
	4 in. or more	1/8 in.
Zone 2	Under 3 in.	1/4 in.
	3 in. or more	1/8 in.
Zone 3	Under 3 in.	1/4 in.
	3 to 6 in.	1/8 in.
	8 in. or more	1/16 in.

15

Vent Installation

Since there are so many types of vents and their role in the plumbing system is so important, many regulations affect the installation of vents. What follows are specifics for installing various vents.

In zone two, any building equipped with plumbing must also be equipped with a main vent. Zone three requires any plumbing system that receives the discharge from a water closet to have either a main vent stack or stack vent. This vent must originate at a 3-in. drainage pipe and extended upward until it penetrates the roof of the building and meets outside air. The vent size requirements for both zones two and three call for a minimum diameter of three in. However, zone two does allow the main stack in detached buildings, where the only plumbing is a washing machine or laundry tub, to have a diameter of 1½ in. Zone one requires all plumbing fixtures, except for exceptions, to be vented.

ROOF PENETRATION

When a vent penetrates a roof, it must be flashed or sealed to prevent water from leaking past the pipe and through the roof. The vent must extend above the roof to a certain height. The height may fluctuate between geographical locations. Average vent extensions are between 12 and 24 in., but check your local regulations to determine the minimum height in your area. Zones one and two generally have height requirements for vent terminations set at 6 in. above the roof. Zone three requires the vent to extend at least 12 in. above the roof.

When vents terminate in the open air, the proximity of their location to windows, doors, or other ventilating openings must be considered. If a vent were placed too close to a window, sewer gas might be drawn into the building when the window was open. Vents should be kept 10 ft from any window, door, opening (such as an attic vent), or ventilation device. If the vent cannot be kept at least 10 ft from the opening, the vent should extend at least 2 ft above it. Zone one requires these vents to extend at least 3 ft above the opening (Fig. 15.1).

FIGURE 15.1   VENT TERMINATION RULES

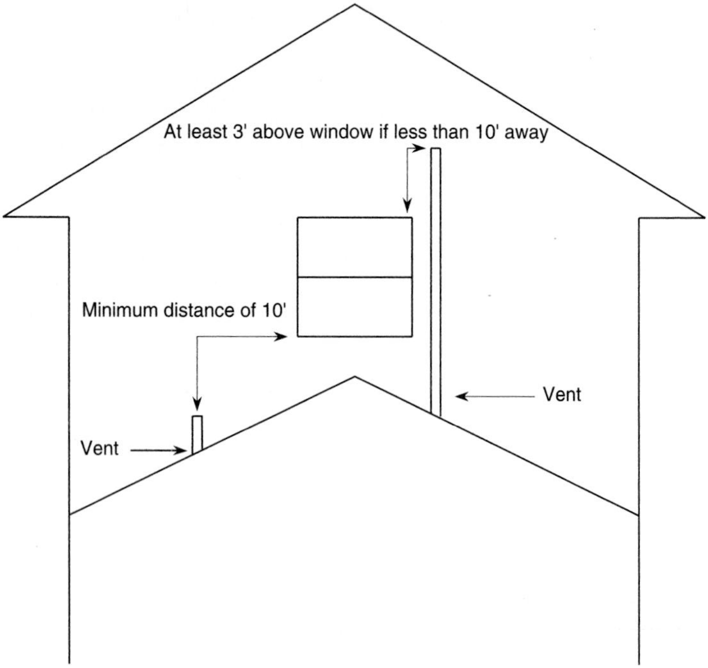

At least 3' above window if less than 10' away

Minimum distance of 10'

Vent

Vent

If the roof being penetrated by a vent is used for activities other than weather protection, such as the roof of a patio, the vent must extend 7 ft above the roof, in zone three. Zone two requires these vents to rise at least 5 ft above the roof.

Cold Climates

In cold climates, vents must be protected from freezing, for condensation can collect on the inside of vent pipes. In cold climates this condensation may turn to ice. As the ice mass grows, the vent becomes blocked and useless. This type of protection is usually accomplished by increasing the size of the vent pipe. This ruling normally applies only in areas where temperatures are expected to be below 0°F. Zone three requires vents in this category to have a minimum diameter of 3 in. If this requires an increase in pipe size, the increase must be made at least 1 ft below the roof. In the case of sidewall vents, the change must be made at least 1 ft inside the wall.

Zone one's rules for protecting vents from frost and snow are a little different. All vents must have diameters of at least 2 in. but never less than the normally required vent size. Any change in pipe size must take place at least 12 in. before the vent penetrates into open air, and the vent must extend to a height of 10 in.

Sidewall Termination

There may be occasions when it is better to terminate a plumbing vent out the side of a wall rather than through a roof. Zone one prohibits sidewall venting. Zone two prohibits sidewall vents from terminating under any building's overhang. When sidewall vents are installed, they must be protected against birds and rodents with a wire mesh or similar cover. Sidewall vents must not extend closer than 10 ft to the boundary of the building lot. If the building is equipped with soffit vents, sidewall vents may not terminate under the soffit vents. This rule is in effect to prevent sewer gas from being sucked into the attic of the house.

Zone three requires buildings having soil stacks with more than five branch intervals to be equipped with a vent stack. Zone one requires a vent stack with buildings having at least 10 stories above the building drain. The vent stack will normally run up near the soil stack. The vent stack must connect into the building drain at or below the lowest branch interval. The vent stack must be sized according to the instructions given earlier. In zone three, the vent stack must be connected within 10 times its pipe size on the downward side of the soil stack. This means that a 3-in. vent stack must be within 30 in. of the soil stack on the downward side of the building drain.

Zone one further requires stack vents to be connected to the drainage stack at intervals of every five stories. The connection must be made with a relief yoke vent. The yoke vent must be at least as large as either the vent stack or soil stack, whichever is smaller. This connection must be made with a wye fitting, at least 42 in. off the floor.

Vent Offsets

In large plumbing jobs, where there are numerous branch intervals, it may be necessary to vent offsets in the soil stack. Normally, the offset must be more than 45° to warrant an offset vent. Zones two and three require offset vents when the soil stack offsets and has five, or more, branch intervals above it.

Grade

Just as drains are installed with a downward pitch, vents must also be installed with a consistent grade. Vents should be graded to allow any water entering the vent pipe to drain into the drainage system. A typical grade for vent piping is 1/4 in./ft. Zone one allows vent pipes to be installed level, without pitch.

Dry Vents

Dry vents must be installed in a manner to prevent clogging and blockages. You may not lay a fitting on its side and use a quarter bend to turn the vent up vertically. Dry vents should leave the drainage pipe in a vertical position. An easy way to remember this is that if you need an elbow to get the vent up from the drainage, you are doing it wrong.

Most vents can be tied into other vents, such as a vent stack or stack vent. But the connection for the tie-in must be at least 6 in. above the flood-level rim of the highest fixture served by the vent.

Circuit Vents

Zone two allows the use of circuit vents to vent fixtures in a battery. The drain serving the battery must be operating at half of its fixture-unit rating. If the application is on a lower-floor battery with a minimum of three fixtures, relief vents are required. You must also pay attention to the fixtures draining above these lower-floor batteries.

When a fixture with a fixture-unit rating of four or less and a maximum drain size of 2 in. is above the battery, every vertical branch must have a continuous vent. If a fixture with a fixture-unit rating exceeding four is present, all fixtures in the battery must be individually vented. Circuit-vented batteries may not receive the drainage from fixtures on a higher level.

Circuit vents should rise vertically from the drainage. However, the vent can be taken off the drainage horizontally if the vent is washed by a fixture with a rating of no more than four fixture-units. The washing cannot come from a water closet. The pipe being washed must be at least as large as the horizontal drainage pipe it is venting.

In zone three, circuit vents may be used to vent up to eight fixtures using a common horizontal drain. Circuit vents must be dry vents, and they should connect to the horizontal drain in front of the last fixture on the branch. The horizontal drain being circuit-vented must not have a grade of more than 1 in./ft. Zone three interprets the horizontal section of drainage being circuit-vented as a vent. If a circuit vent is venting a drain with more than four water closets attached to it, a relief vent must be installed in conjunction with the circuit vent.

Vent Placement

Vent placement in relation to the trap it serves is important and regulated. The maximum allowable distance between a trap and its vent will depend on the size of the fixture drain and trap. All vents, except those for fixtures with integral traps, should connect above the trap seal. A sanitary tee fitting should be used when going from a vertical stack vent to a trap. Other fittings with a longer turn, such as a combination wye and eighth bend, will place the trap in more danger of back-siphonage. I know this goes against the common sense of a smoother flow of water, but the sanitary tee reduces the risk of a vacuum.

SUPPORTING VENT PIPES

Vent pipes must be supported, and they may not be used to support antennas, flagpoles, and similar items. Depending upon the type of material you use and whether the pipe is installed horizontally or vertically, the spacing between hangers will vary. Both horizontal and vertical pipes require support. The regulations in the plumbing code apply to the maximum distance between hangers.

SOME MORE VENTING REGULATIONS FOR ZONE ONE

Some interceptors, like those used as a settling tank that discharges through a horizontal indirect waste, are not required to be vented. However, the interceptor receiving the discharge from the unvented interceptor must be properly vented and trapped.

Traps for sinks that are a part of a piece of equipment, like a soda fountain, are not required to be vented when venting is impossible. But these drains must drain through an indirect-waste to an approved receptor.

OTHER VENTING REQUIREMENTS FOR ZONE TWO

All soil stacks that receive the waste of at least two vented branches must be equipped with a stack vent or a main stack vent. Except when approved, fixture drainage is not allowed to enter a stack at a point above a vent connection. Side-inlet closet bends are allowed to accept the connection of fixtures that are vented. However, these connections may not be used to vent a bathroom unless the connection is washed by a fixture. All fixtures dumping into a stack below a higher fixture must be vented, except when special approval is granted for a variance. Stack vents and vent stacks must connect to a common vent header before vent termination.

Traps for sinks that are a part of a piece of equipment, like a soda fountain, are not required to be vented when venting is impossible. But these drains must be piped in accordance with the combination waste and vent regulations for zone two.

Up to two fixtures, set back-to-back or side-by-side, within the allowable distance between the traps and their vents may be connected to a common horizontal branch that is vented by a common vertical vent. However, the horizontal branch must be one pipe-size larger than normal. When applying this rule, the following rating applies: shower drains, 3-in. floor drains, 4-in. floor drains, pedestal urinals, and water closets with fixture-unit ratings of four fixture-units shall be considered to have 3-in. drains.

Some fixture groups are allowed to be stack vented without individual back vents. These fixture groups must be located in one-story buildings or on the top floor of the building, with some special provisions. Fixtures located on the top floor must connect independently to the soil stack, and the bathing units and water closets must enter the stack at the same level.

This same stack-venting procedure can be adapted to work with fixtures on lower floors. The stack being stack-vented must enter the main soil stack though a vertical eighth bend and wye combination. The drainage must enter above the eighth bend. A 2-in. vent must be installed on the fixture group. This vent must be 6 in. above the flood-level rim of the highest fixture in the group.

Some fixtures are allowed to be served by a horizontal waste that is within a certain distance to a vent. When piped in this manner, bathtubs and showers are both required to have 2-in. P-traps. These drains must run with a minimum grade of 1/4 in./ft. A single drinking fountain can be rated as a lavatory for this type of piping. On this type of system, fixture drains for lavatories may not exceed 1¼ in., and sink drains cannot be larger than 1½ in. in diameter.

In multistory situations, it is possible to drain up to three fixtures into a soil stack above the highest water closet or bathtub connection, without re-venting. To do this, the following requirements must be met:

Minimum stack size of 3 in. is required.

Approved fixture-unit load on stack is required.

All lower fixtures must be properly vented.

All individually unvented fixtures must be within allowable distances to the main vent.

Fixture openings must not exceed the size of their traps.

All code requirements must be met and approved.

Fixtures that are allowed to be stack-vented without indvidual vents in zone two are the following:

Water closets

Basins

Bathtubs

Showers

Kitchen sinks, with or without dishwasher and garbage disposer

Restrictions apply to this type of installation.

TRAP-TO-VENT DISTANCES IN ZONE ONE

Grade on Drain Pipe (inches)	Size of Trap Arm (inches)	Maximum Distance Between Trap and Vent
1/4	1¼	2 ft 6 in.
1/4	1½	3 ft 6 in.
1/4	2	5 ft
1/4	3	6
1/4	4 and larger	10 ft

TRAP-TO-VENT DISTANCES IN ZONES TWO AND THREE

	Grade on Drain Pipe (inches)	Fixture's Drain Size (inches)	Trap Size (inches)	Maximum Distance Between Trap and Vent
Zone Two	1/4	1¼	1¼	3 ft 6 in.
	1/4	1½	1¼	5 ft
	1/4	1½	1½	5 ft
	1/4	2	1½	8 ft
	1/4	2	2	6 ft
	1/8	3	3	10 ft
	1/8	4	4	12 ft
Zone Three	1/4	1¼	1¼	3 ft 6 in.
	1/4	1½	1¼	5 ft
	1/4	1½	1½	5 ft
	1/4	2	1½	8 ft
	1/4	2	2	6 ft
	1/8	3	3	10 ft
	1/8	4	4	12 ft

VENT SIZE

Type of Fixture	Minimum Size of Vent (inches)
Lavatory	1¼
Drinking fountain	1¼
Domestic sink	1¼
Shower stalls, domestic	1¼
Bathtub	1¼
Laundry tray	1¼
Service sink	1¼
Water closet	2

Note: At least 1 3-in. vent must be installed.

16

Gas Piping

The installation of gas piping is serious business. Depending upon where you work, the requirements pertaining to gas work may or may not be set forth and enforced by the plumbing code. Many jurisdictions mingle gas regulations with plumbing rules. For this reason, we will go over many of the criteria common to gas work where it is dealt with by licensed plumbers.

APPROVED MATERIALS FOR WORKING WITH GAS

Several types of piping materials are approved for gas work. All piping used must meet minimum requirements, as established by local codes. The two materials most often used for gas piping in buildings are steel and copper. When copper is used, it should be either type L or type K, and it must be approved for use with gas. Pipes made of PVC and PE are usually allowed for gas pipe in buried installations outside a building.

Metallic pipe can be used in buildings and above ground so long as the gas being conveyed will not corrode the pipe. Steel pipe, approved copper pipe, and yellow brass pipe are the three types of pipes required for use in zone one. Aluminum pipe, where it is approved, must not be used below ground. It must not be used outside, and when used inside, it must not come into contact with masonry, plaster, or insulation. Furthermore, it must be protected from contact with moisture.

Ductile iron pipe, when approved, is allowed only for underground use, outside of buildings. If any pipe could be subject to corrosive action from surrounding conditions, it must be protected to avoid corrosive action.

The fittings used with gas pipe must be compatible with the pipe. They must also be approved fittings. In gas work, bushings are not generally allowed. Increasers and reducers are normally acceptable. Zone one allows the use of bushings if they are not concealed.

Flexible connectors are often used to connect an appliance to a gas source. These connectors must be approved and marked to prove it. Flex connectors may not be longer than 6 ft. Zone one requires appliance connectors for all appliances, except ranges and dryers, to be no more than 3 ft long. Flex connectors may not be concealed in walls, floors, or partitions. Furthermore, flex connectors may not penetrate walls, floors, or partitions. Flex connectors must be properly sized. They may not be smaller than the inlet of the device they are serving.

Gas hose is not a flex connector. Gas hose is generally prohibited, except for special circumstances. Such circumstances could include a biology lab, where gas burners need to be moved around. If gas hose is approved for use, it must be as short as reasonably possible, and it may not exceed 6 ft in length. This length restriction does not apply to items like handheld torches.

Gas hose may not be concealed, and it may not penetrate walls, floors, or partitions. If the hose will be exposed to temperatures above 125°F, it may not be used. If allowed for use, gas hose must be connected to a cutoff valve at the gas pipe supply. This type of hose may be used on outdoor appliances that are designed to be portable. In these uses, the length of the hose may not exceed 15 ft. The hose still must connect to a cutoff valve at the gas pipe supply.

When flex connectors are not used, soft copper tubing often is. In zone one, quick-disconnect connectors are approved. These devices allow the connection to be broken by hand, and the gas is shut off automatically. When copper is used at an appliance connection, it should be type L or type K, and it must not be bent in a manner to damage the structural qualities of the tubing. All pipe bending must be done with approved equipment.

INSTALLING GAS PIPE

Installing gas pipe is not the same as installing plumbing pipes. There are similarities, but the procedure is not the same. One difference is in the way pipe and fittings are put together. All joints must be made gas-tight. The joints should be tested with a mercury gauge, at the required pressure, to ensure good joints. Zone one requires the test to maintain 6 in. of mercury. If the test uses a pressure gauge, the test must maintain 10 lb/in.^{2} Air is commonly used to provide the pressure test on gas pipe. The pipe must maintain its test pressure for at least 15 min.

In pipes carrying gas at high pressure, the test pressure is required to be 60 lb/in.^{2} These high-pressure tests are often required to be maintained for 30 min.

All pipe ends are to be cut squarely and with a full diameter. Any burrs on the pipe must be removed. The surfaces of a gas joint must be clean. If flux is used to make a joint, the flux must be approved for the purpose. When threaded pipe is used, only the male threads are allowed to be sealed with pipe dope or tape. Mechanical joints must be used according to the manufacturer's specification.

The only two types of pipes allowed to have heat-fusion joints are PE and PB, where approved. However, these two types of pipe may not be used with a cement or glue joint. When PVC pipe is used, it must be primed and glued with approved materials.

When more than one type of piping is used, the joint between the opposing pipe types must be made with an approved adapter fitting. In the case of matching metallic pipes together, a dielectric fitting is generally required.

In general, all underground gas piping must be installed at a depth of at least 18 in. Zone one requires buried metallic pipe to be covered by a minimum of 12 in. of dirt. The pipe must not be installed in a way to hinder maintenance or to place the pipe in jeopardy of damage. There is, of course, an exception to this rule. Many states allow gas lines serving an individual outside appliance to be buried 8 in. deep, but zone one does not recognize this exception. This exception, as usual, is subject to local inspection and approval.

Any underground gas pipe penetrating a foundation must be protected by a pipe sleeve. A rule-of-thumb sizing for the sleeve is two pipe-sizes larger than the gas pipe. The additional space in the sleeve must be sealed to prevent water, insect, or vermin invasion. Just as with plumbing, gas pipe located in flood areas must be protected against flooding and the complications associated with flooding.

Piping for gas, other than dry gas, must be graded with a pitch of a 1/4-in. fall for every 15 ft the pipe runs. At any point where the pipe is low or condensation may occur, a drip leg is required. Drip legs must be accessible and must be protected from freezing temperatures.

The connection of branch piping to a main distribution pipe must be made either on top of the main or on the side but not on the bottom. Bottom connections are not the only prohibited practices. Gas pipe may not be installed in or through heat ducts, air ducts, chimneys, laundry drops, vents, dumbwaiters, or elevator shafts.

Concealed piping may not have union connections. Tubing fittings and running threads are also prohibited in concealed locations. There is yet another rule pertaining to concealed gas piping. Unless the pipe is made of steel, it must be protected from punctures. This is most easily accomplished with the use of nail plates. Nail plates are required when a pipe other than steel is positioned within 1¼ in. from the surface of a wood member, such as a stud or floor joist. The nail plate must have a minimum thickness of 1/16 in. The plate must be large enough to protect the pipe from punctures, which usually means installing a plate that extends at least 4 in. beyond a normal nailing surface.

When installing gas pipe in concrete, you must observe still more regulations. Gas pipe buried in concrete
must be covered by no less than 1½ in. of concrete. The pipe
must not make contact with metal objects, and the concrete
must not contain materials that will have an adverse affect
on the piping.

SUPPORTS

Gas pipe can get heavy, and the support for the piping
must be capable of carrying this weight. Hangers and supports should be made of approved materials that are intended for use with the type of pipe being supported. The
required intervals of support are different from those in
plumbing.

All gas pipe installed above ground must be supported in an approved manner and protected from damage.
Zone one's requirements for support are as follows: 1/2-in.
pipe must be supported at intervals not to exceed 6 ft; 3/4-
in. and 1-in. pipe at 8-ft intervals; and larger pipe at 10-ft
intervals. Pipe with a diameter of at least 1¼ in. is required to be supported at each floor level when installed
vertically.

The intervals for pipe support in other zones also depend upon the type of pipe used and the size of the pipe. For
example, when supporting tubing with a diameter of 1½ in.
or more, support must be present every 10 ft. This is the
same distance allowed for supports holding rigid pipe with
a diameter of 3/4 in. or less. Rigid pipe with a diameter of 1
in. or larger needs to be supported only every 12 ft. Smaller
tubing with a diameter of 1¼ in. or less requires support at
minimum intervals of 6 ft.

ROUTINE REGULATIONS

There are some routine regulations that you should know
about gas piping. Every building housing gas piping must
have a cutoff valve located on the outside of the building.
This is a big help in the event of a fire. All gas meters are required to be equipped with shutoffs on the incoming side of
the meter. Cutoff valves are required at all locations where

appliances connect to gas supply pipes. These valves must not only be accessible, they must be adjacent to the appliance. Of course, all cutoffs must be of an approved type. The connection between an appliance and a supply pipe must be equipped with an approved union fitting. If an appliance is removed or a gas pipe is not in use, the pipe must be capped to prevent any gas from escaping.

When gas is provided by a bulk dispenser, the dispenser must have an emergency cutoff switch. A backflow preventer must be installed on the supply side of the dispenser. All gas-dispensing systems that are located inside a building must be vented in an approved fashion. Back-flow preventers are also required on systems using a backup gas source or a supplemental gas source.

ADDITIONAL ZONE ONE REQUIREMENTS FOR LIQUID PETROLEUM (LP) GAS

The relief valves for LP gas must discharge into the open air. These valves must not be located closer than 5 ft, measured horizontally, from any opening into a building. Liquid petroleum gas may not be piped to water heaters in locations where gas might collect and provide opportunity for fire or explosion.

CONCEALED PIPING

All concealed gas piping must be tested and approved before it is concealed. A standard test pressure is a pressure equal to 1½ times the normal working pressure of the system. However, the test pressure must never be less than 3 psi. With LP gas, the test pressure must equate to an 18-in. water column. The test must be conducted for a minimum of 10 min. To be approved, the system must not lose pressure during the test. A mercury gauge is the most common way of testing gas pipe. If the piping loses pressure, leaks should be located with soapy water, not fire or acid. When leaks are located, defective pipe or fittings should be removed and replaced, not repaired.

Once the test is done, the only job left is to purge the system and to get it on line. It is not permissible to purge the gas system through an appliance. The purging must be done in a safe location, where combustion is not a potential threat.

SIZING

Your local gas code contains tables for use in selecting pipe of the proper size. These tables are based on a few factors, which include maximum capacity of the pipe, gas pressure, pressure drop, gravity, and pipe length. The sizing tables provided with gas codes are illustrated and described to make sizing relatively easy.

Most buildings are restricted to a maximum operating pressure of 5 psig. There are exceptions to this rating, but 5 psig is an average rating and is based on the gas in the pipe being natural gas. If the gas is propane, the numbers change. Liquid petroleum (LP) gas is meant for a maximum operating pressure of 20 psig. As usual, there will be exceptions to this rule, but 20 psig is normal.

REGULATORS

Regulators are often needed to regulate gas pressure. If a regulator is used outside, it must be approved for exterior use. Some regulators require an individual vent. When such a vent is required, it must be piped independently to the outside of the building. The vent must be protected against damage and the influx of foreign objects.

Gas regulators must be installed in accessible locations. All regulators must be installed in a manner to prevent them from being damaged. A regulator is required when a gas appliance is designed to work at a lower gas pressure than the pressure present in the piping. If a second-stage regulator is required for LP gas, it must be an approved model.

SOME MORE REGULATIONS FOR ZONE ONE

Zone one has more gas regulations. Used pipe, unless it was used for gas, may not be used in gas installations. If gas pipe is welded, it must be welded by a certified pipeline welder.

Exposed gas pipe must be installed in a way to keep it at least 6 in. above the ground or other obstructions. It is not permissible to install gas piping below grade within the confines of a building. When special cavities are provided, gas pipe may be concealed and unprotected.

Underground ferrous gas piping must be protected from electricity with isolation fittings that are installed at least 6 in. above ground. If unions are installed in gas pipe, they must be installed with right and left nipples and couplings. Exposed unions may be used. When gas pipe serves multiple buildings or tenants, there must be an individual cutoff valve installed for each user. These valves must be installed outside, and they must be readily accessible at all times.

When more than one type of gas has access to a gas pipe, the pipe must be protected against back-flow. Gas-fired barbecues and fireplaces must be controlled with approved valves. The valves must be in the same room as the gas-fired unit. However, the valve may not be in the unit or on a hearth that serves the unit. These valves must be installed within 4 ft of the gas outlet for the gas-fired unit. The pipe going from the valve to the unit may be installed in concrete or masonry if the pipe is a standard weight brass or galvanized steel and if there will be at least 2 in. of concrete or masonry around the pipe.

Cutoffs for gas appliances are required to be within 3 ft of the appliance. The cutoffs must be of an approved type. They must be installed on the gas supply pipe. They must be installed in front of unions that are between the gas supply pipe and an appliance. Cutoffs can be placed adjacent to, in, or under appliances so long as the appliance can be moved, without affecting the cutoff. When cutoffs are installed in or under gas-fired units, they must be accessible. Appliances may not be piped in a way to allow a gas supply from more than one gas piping system.

When installing underground gas pipe that is not metallic, you must also install a No. 18 copper wire. The wire must run with and be attached to the gas pipe. The wire must be exposed above grade at both ends of the pipe run.

PIPE SIZING CALCULATIONS

Pipe Capacity (cubic feet of gas per hour)[a]

Tubing Diameter Inside (Outside) (inches)	5	10	15	20	30	40	50	60	70	80	90	100
					Length of Tubing (feet)							
1/4 (3/8)	540	360	285	240	192	163	143	130	118	110	102	96
3/8 (1/2)	1,260	850	670	570	450	380	335	300	275	255	240	225
1/2 (5/8)	2,400	1,630	1,280	1,080	860	730	645	580	530	490	460	430
5/8 (3/4)	4,150	2,780	2,150	1,860	1,480	1,250	1,100	1,000	910	850	790	740
3/4 (7/8)	6,500	4,350	3,450	2,950	2,300	2,000	1,750	1,560	1,430	1,330	1,230	1,160
1 (1⅛)	10,500	7,600	6,200	5,400	4,400	3,800	3,350	3,050	2,800	2,650	2,500	2,350
1¼ (1⅜)	21,000	15,000	12,000	10,500	8,600	7,500	6,700	6,100	5,600	5,200	4,900	4,700
1½ (1⅝)	31,000	22,000	18,000	15,000	13,000	11,000	9,800	9,000	8,200	7,700	7,200	6,800
2 (2⅛)	58,000	41,000	34,000	29,000	24,000	20,000	18,000	17,000	15,000	14,000	13,500	12,500
2½ (2⅝)	90,000	64,000	52,000	45,000	36,000	32,000	28,000	26,000	24,000	22,000	21,000	20,000
3 (3⅛)	150M	110M	90,000	79,000	63,000	55,000	49,000	45,000	41,000	38,000	36,000	34,500
4 (4⅛)	310M	220M	180M	150M	125M	110M	97,000	90,000	82,000	77,000	72,000	70,000

[a] 2 lb pressure capacity of pipes or tubing from point of delivery to 2 psig regulator, based on a pressure drop of 1.5 psi and a gas of 0.65 specific gravity.

PIPE SIZING CALCULATIONS (cont.)

Tubing Diameter Inside (Outside) (inches)	Pipe Capacity (cubic feet of gas per hour)[a] Length of Tubing (feet)											
	5	10	15	20	30	40	50	60	70	80	90	100
1/4 (3/8)	62	42	33	28	22	19	16	15	14	13	12	11
3/8 (1/2)	145	96	76	66	52	44	39	35	32	30	28	26
1/2 (5/8)	280	187	148	126	100	84	74	67	61	57	53	50
5/8 (3/4)	475	320	252	215	170	145	129	115	105	97	91	86
3/4 (7/8)	750	500	395	335	265	225	200	180	165	153	142	134
1 (1⅛)	1,080	760	620	520	440	380	345	315	290	270	255	240
1¼ (1⅜)	2,200	1,550	1,300	1,100	900	795	700	640	600	560	520	500
1½ (1⅝)	3,450	2,450	2,000	1,700	1,400	1,200	1,100	1,000	920	860	810	770
2 (2⅛)	7,000	4,950	4,000	3,450	2,850	2,500	2,200	2,000	1,850	1,750	1,650	1,580
2½ (2⅝)	11,200	8,000	6,500	5,600	4,600	4,000	3,550	3,250	3,000	2,800	2,650	2,500
3 (3⅛)	20,000	14,200	11,500	10,000	8,300	7,200	6,350	5,800	5,400	5,050	4,750	4,500
4 (4⅛)	42,000	29,500	24,000	21,000	17,000	14,500	13,000	12,000	11,000	10,400	9,700	9,300

[a]2 lb pressure capacity of pipes or tubing from 2 psig regulator to appliance, based on a pressure drop of 1 in. water column and a gas of 0.65 specific gravity.

PIPE SIZING CALCULATIONS (cont.)

Pipe Size of Schedule 40 Standard Pipe (inches)	Internal Diameter (inches)	Pipe Capacity (cubic feet of gas per hour)[a]								
		Length of Pipe (feet)								
		50	100	150	200	250	300	400	500	1,000
1	1.049	1,989	1,367	1,098	940	833	755	646	572	393
1¼	1.380	4,084	2,807	2,254	1,929	1,710	1,549	1,326	1,175	808
1½	1.610	6,120	4,206	3,378	2,891	2,562	2,321	1,987	1,761	1,210
2	2.067	11,786	8,101	6,505	5,567	4,934	4,471	3,827	3,391	2,331
2½	2.469	18,785	12,911	10,368	8,874	7,865	7,126	6,099	5,405	3,715
3	3.068	32,209	22,824	18,329	15,687	13,903	12,597	10,782	9,556	6,568
3½	3.548	48,623	33,418	26,836	22,968	20,356	18,444	15,786	13,991	9,616
4	4.026	67,736	46,555	37,385	31,997	28,358	25,694	21,991	19,490	13,396
5	5.047	122,544	84,224	67,635	57,887	51,304	46,485	39,785	35,261	24,235
6	6.065	198,427	136,378	109,516	93,732	83,073	75,270	64,421	57,095	39,241
8	7.981	407,692	280,204	225,014	192,583	170,683	154,651	132,361	117,309	80,626
10	10.020	740,477	508,926	408,686	349,782	310,005	280,887	240,403	213,065	146,438
12	11.938	1,172,269	805,694	647,001	553,749	490,777	444,680	380,588	337,309	231,830

[a] 5 lb pressure capacity of pipes for an initial pressure of 5 psig with a 10 percent pressure drop and a gas of 0.0 specific gravity.

PIPE SIZING CALCULATIONS (cont.)

Pipe Size of Schedule 40 Standard Pipe (inches)	Internal Diameter (inches)	Pipe Capacity (cubic feet of gas per hour)[a]									
		Length of Pipe (feet)									
		10	20	30	40	50	60	70	80	90	100
1/4	.364	43	29	24	20	18	16	15	14	13	12
3/8	.493	95	65	52	45	40	36	33	31	29	27
1/2	.622	175	120	97	82	73	66	61	57	53	50
3/4	.824	360	250	200	170	151	138	125	118	110	103
1	1.049	680	465	375	320	285	260	240	220	205	195
1¼	1.380	1,400	950	770	660	580	530	490	460	430	400
1½	1.610	2,100	1,460	1,180	990	900	810	750	690	650	620
2	2.067	3,950	2,750	2,200	1,900	1,680	1,520	1,400	1,300	1,220	1,150

[a]Maximum capacity of pipe for gas pressure of 1/2 psig or less with pressure drop of 1/2-in. water column and 0.60 specific gravity.

PIPE SIZING CALCULATIONS (cont.)

Pipe Size of Schedule 40 Standard Pipe (inches)	Internal Diameter (inches)	Pipe Capacity (cubic feet of gas per hour)[a]									
		Length of Pipe (feet)									
		10	20	30	40	50	60	70	80	90	100
1/4	.364	32	22	18	15	14	12	11	11	10	9
3/8	.493	72	49	40	34	30	27	25	23	22	21
1/2	.622	132	92	73	63	56	50	46	43	40	38
3/4	.824	278	190	152	130	115	105	96	90	84	79
1	1.049	520	350	285	245	215	195	180	170	160	150
1¼	1.380	1,050	730	590	500	440	400	370	350	320	305
1½	1.610	1,600	1,100	890	760	670	610	560	530	490	460
2	2.067	3,050	2,100	1,650	1,450	1,270	1,150	1,050	990	930	870

[a]Maximum capacity of pipe for gas pressure of 1/2 psig or less with pressure drop of 3/10-in. water column and 0.60 specific gravity.

PIPE SIZING CALCULATIONS (cont.)

Tubing Capacity (cubic feet of gas per hour)[a]

Outside Diameter (inch)	10	20	30	40	50	60	70	80	90	100
					Length of Tubing (feet)					
3/8	20	14	11	10	9	8	7	7	6	6
1/2	42	29	23	20	18	16	15	14	13	12
5/8	86	59	47	40	36	33	30	28	26	25
3/4	150	103	83	71	63	57	52	49	46	43
7/8	212	146	117	100	89	81	74	69	65	61

[a]Maximum capacity of semirigid tubing for gas pressure of 1/2 psig or less with pressure drop of 3/10-in. water column and 0.60 specific gravity.

PIPE SIZING CALCULATIONS *(cont.)*

Outside Diameter (inch)	Tubing Capacity (cubic feet of gas per hour)[a]									
	Length of Tubing (feet)									
	10	20	30	40	50	60	70	80	90	100
3/8	27	18	15	13	11	10	9	9	8	8
1/2	56	38	31	26	23	21	19	18	17	16
5/8	113	78	62	53	47	43	39	37	34	33
3/4	197	136	109	93	83	75	69	64	60	57
7/8	280	193	155	132	117	106	98	91	85	81

[a]Maximum capacity of semirigid tubing for gas pressure of 1/2 psig or less with pressure drop of 1/2-in. water column and 0.60 specific gravity.

PIPE SIZING CALCULATIONS (cont.)

Pipe Size of Schedule 40 Standard Pipe (inches)	Internal Diameter (inches)	Pipe Capacity (cubic feet of gas per hour)[a]										
		Total Equivalent Length of Pipe (feet)										
		50	100	150	200	250	300	400	500	1,000
1	1.049	215	148	119	102	90	82	70	62	43
1¼	1.380	442	304	244	209	185	168	143	127	87
1½	1.610	622	455	366	313	277	251	215	191	131
2	2.067	1,275	877	704	602	534	484	414	367	252
2½	2.469	2,033	1,397	1,122	960	851	771	660	585	402
3	3.068	3,594	2,470	1,983	1,698	1,505	1,363	1,167	1,034	711
3½	3.548	5,262	3,616	2,904	2,485	2,203	1,996	1,708	1,514	1,041
4	4.026	7,330	5,038	4,046	3,462	3,069	2,780	2,380	2,109	1,450
5	5.047	13,261	9,114	7,319	6,264	5,552	5,030	4,305	3,816	2,623
6	6.065	21,472	14,758	11,851	10,143	8,990	8,145	6,971	6,178	4,246
8	7.981	44,118	30,322	24,350	20,840	18,470	16,735	14,323	12,694	8,725
10	10.020	80,130	55,073	44,225	37,851	33,547	30,396	26,015	23,056	15,847
12	11.938	126,855	87,187	70,014	59,923	53,109	48,120	41,185	36,501	25,087

[a]Pipe sizing table for pressures under 1 lb approximate capacity of pipes with pressure drop of 3/10-in. water column and 0.60 specific gravity.

PIPE SIZING CALCULATIONS (cont.)

Pipe Size of Schedule 40 Standard Pipe (inches)	Internal Diameter (inches)	Pipe Capacity (cubic feet of gas per hour)[a]								
		Total Equivalent Length of Pipe (feet)								
		50	100	150	200	250	300	400	500	1,000
1	1.049	284	195	157	134	119	108	92	82	56
1¼	1.380	583	400	322	275	244	221	189	168	115
1½	1.610	873	600	482	412	366	331	283	251	173
2	2.067	1,681	1,156	928	794	704	638	546	484	333
2½	2.469	2,680	1,842	1,479	1,266	1,122	1,017	870	771	530
3	3.068	4,738	3,256	2,615	2,238	1,983	1,797	1,538	1,363	937
3½	3.548	6,937	4,767	3,828	3,277	2,904	2,631	2,252	1,996	1,372
4	4.026	9,663	6,641	5,333	4,565	4,046	3,666	3,137	2,780	1,911
5	5.047	17,482	12,015	9,649	8,258	7,319	6,632	5,676	5,030	3,457
6	6.065	28,308	19,456	15,624	13,372	11,851	10,738	9,190	8,145	5,598
8	7.981	58,161	39,974	32,100	27,474	24,350	22,062	18,883	16,735	11,502
10	10.020	105,636	72,603	58,303	49,900	44,225	40,071	34,296	30,396	20,891
12	11.938	167,236	114,940	92,301	78,998	70,014	63,438	54,295	48,120	33,073

[a]Pipe sizing table for pressures under 1 lb approximate capacity of pipes with pressure drop of 1/2-in. water column and 0.60 specific gravity.

PIPE SIZING CALCULATIONS *(cont.)*

Pipe Size of Schedule 40 Standard Pipe (inches)	Internal Diameter (inches)	Pipe Capacity (cubic feet of gas per hour)[a]									
		Total Equivalent Length of Pipe (feet)									
		50	100	150	200	250	300	400	500	1,000
1	1.049	717	493	396	338	300	272	233	206	142
1¼	1.380	1,471	1,011	812	695	616	558	478	423	291
1½	1.610	2,204	1,515	1,217	1,041	923	836	716	634	436
2	2.067	4,245	2,918	2,343	2,005	1,777	1,610	1,378	1,222	840
2½	2.469	6,766	4,651	3,735	3,196	2,833	2,567	2,197	1,947	1,338
3	3.068	11,962	8,221	6,602	5,650	5,008	4,538	3,884	3,442	2,366
3½	3.548	17,514	12,037	9,666	8,273	7,332	6,644	5,686	5,039	3,464
4	4.026	24,398	16,769	13,466	11,525	10,214	9,255	7,921	7,020	4,825
5	5.047	44,140	30,337	24,362	20,851	18,479	16,744	14,330	12,701	8,729
6	6.065	71,473	49,123	39,447	33,762	29,923	27,112	23,204	20,566	14,135
8	7.981	146,849	100,929	81,049	69,368	61,479	55,705	47,676	42,254	29,041
10	10.020	266,718	183,314	147,207	125,990	111,663	101,175	86,592	76,745	52,747
12	11.938	422,248	290,209	233,048	199,459	176,777	160,172	137,087	121,498	83,505

[a]Pipe sizing table for pressures under 1 lb pressure capacity of pipes for an initial pressure of 1 psig with a 10 percent pressure drop and a gas of 0.60 specific gravity.

PIPE SIZING CALCULATIONS (cont.)

Pipe Size of Schedule 40 Standard Pipe (inches)	Internal Diameter (inches)	Pipe Capacity (cubic feet of gas per hour)[a]								
		Total Equivalent Length of Pipe (feet)								
		50	100	150	200	250	300	400	500	1,000
1	1.049	1,112	764	614	525	466	422	361	320	220
1¼	1.380	2,283	1,569	1,260	1,079	956	866	741	657	452
1½	1.610	3,421	2,351	1,888	1,616	1,432	1,298	1,111	984	677
2	2.067	6,589	4,528	3,636	3,112	2,758	2,499	2,139	1,896	1,303
2½	2.469	10,501	7,217	5,796	4,961	4,396	3,938	3,409	3,022	2,077
3	3.068	18,564	12,759	10,246	8,769	7,772	7,042	6,027	5,342	3,671
3½	3.548	27,181	18,681	15,002	12,840	11,379	10,311	8,825	7,821	5,375
4	4.026	37,865	26,025	20,899	17,887	15,853	14,364	12,293	10,895	7,488
5	5.047	68,504	47,082	37,809	32,359	28,680	25,986	22,240	19,711	13,547
6	6.065	110,924	76,237	61,221	52,397	46,439	42,077	36,012	31,917	21,936
8	7.981	227,906	156,638	125,786	107,657	95,414	86,452	73,992	65,578	45,071
10	10.020	413,937	284,497	228,461	195,533	173,297	157,020	134,389	119,106	81,861
12	11.938	655,315	450,394	361,682	309,553	274,351	248,582	212,754	188,560	129,596

[a]Pipe sizing table for 2 lb pressure capacity of pipes for an initial pressure of 2 psig with a 10 percent pressure drop and a gas of 0.60 specific gravity.

17

Testing Systems

Testing systems once they have been installed is part of the plumbing process. Code regulations govern when and how tests are to be done. To comply with your local plumbing code, you must be aware of these rules and regulations.

TESTING SEWERS

There are two common methods for testing building sewers. The first method uses water, the second air. In either case, the building sewer should be capped or plugged at the point at which it connects with the main sewer. Test tee fittings are commonly installed in this portion of the sewer to allow for the test. Sewers must be tested, inspected, and approved before they are covered. Sewers should be covered by a minimum of 12 in. of earth.

When testing with water, fill the sewer with water to a point equal to a 10-ft head. In simple terms, this means extending a pipe, like a clean-out riser, to a point 10 ft higher than the sewer. The pipe rising to allow for the 10-ft head should have water resting at its upper limit. The water must be visible.

When testing with water, water pressure must be maintained for at least 15 min before an official inspection is made. If the water level goes down, you've got a problem. All joints must be watertight. If leaks are present, cut out the defective section and replace it. Don't attempt to patch leaks with temporary measures.

When testing with air, you must rig the sewer to accept a pressure gauge. The sewer must be pumped with air until the contents reach a pressure of at least 5 psi. If a mercury gauge is used, the pressure must balance 10 in. of mercury. The time requirements for an air test are the same as those for a water test.

TESTING A WATER SERVICE

The test of a water service is sometimes waived. If the water service is comprised of a single pipe with no joints, a pressure test may not be required. If a test is required, the pipe can be tested with potable water or air. The water service must be tested (if a test is required), inspected, and approved before being buried. Water services must be located deep enough to prevent freezing.

Zones one and three require water services to be tested at a pressure equal to their maximum working pressure. Zone two requires the test pressure to be set at a pressure of at least 25 psi higher than the maximum working pressure.

TESTING GROUNDWORKS

Underground plumbing is tested in essentially the same way that a building sewer is tested. An air pressure of 5 psi or a 10-ft head of water is required. When a mercury gauge is used, the test must balance a 10-in. column of mercury. The test must be maintained for at least 15 min prior to inspection.

TESTING DWV ROUGH-INS

All DWV rough-ins must be inspected before being concealed. When testing with air, vent terminals, fixture outlets, and the building drain must be capped or plugged. The DWV system must be subjected to a 15-min test with either air or water. If air is used, the system must be tested with a minimum pressure of 5 psi or a 10-in. column of mercury.

When a DWV system is tested with water, the water level is usually required to extend to the top of the roof vents. Some areas allow the test to terminate at the flood-level rim of the highest bathing unit in the premises, but normally the water must be to the top of the vents.

During the DWV test, inspectors look for pipe protection. When a pipe is installed in a way that may allow it to be penetrated by nails or screws, it must be protected with nail plates. If structural members have been substantially weakened by your plumbing installation, your job will not pass inspection. Pipe hangers are also inspected.

TESTING WATER DISTRIBUTION ROUGH-INS

All water distribution pipes must be tested, inspected, and approved before being concealed. The test pressure required for a potable water system is usually the same as the maximum working pressure for the system. However, zone two requires the test pressure to be 25 psi higher than the working pressure.

Another consideration in inspection of water piping includes pipe protection from punctures and freezing. Pipe hangers are also inspected. Back-flow preventers, air gaps, and all other code requirements are examined in these inspections, as they are with other inspections.

TESTING FINAL PLUMBING

The job is not done until the final approval is issued from the code enforcement office. What is involved in a final plumbing inspection? Well, typically, the inspection is a matter of a visual tour of the plumbing. This tour normally includes the inspector's use and observation of all plumbing fixtures. For example, an inspector tests to see that the hot water is piped to the left side of a faucet. The inspector checks traps and other connections for leaks.

In the final inspection, inspectors put all the plumbing fixtures through their paces. Cutoffs are inspected, aerators are checked, back-flow preventers are checked, fixtures are filled and drained, and water heaters are tested. In general, all plumbing is checked to assure proper installation procedures and working conditions.

If an inspector has reason to suspect a plumbing system is not up to snuff, the inspector may require a smoke or peppermint test. These tests are designed to expose leaks in the DWV system. In these tests, all traps are filled with water. The DWV system is filled with a colored smoke or oil of peppermint. When the smoke is visible at a vent or the peppermint odor is noticeable, the vents are capped. Then the inspector checks each trap for evidence of a leak. The colored smoke or aromatic peppermint make it easy to find traps that are not doing their job.

INTERIOR RAIN LEADERS AND DOWNSPOUTS

Interior rain leaders and downspouts should be tested, inspected, and approved in the same manner used for DWV systems. These pipes should not be concealed before testing, inspecting, and approval.

18

Blueprint Symbols

This chapter is comprised of common symbols used to reference plumbing fixtures on blueprints. Take a few moments to study the illustrations. If you will be working from blueprints, as many plumbers do, these symbols are likely to show up. Learn the symbols ahead of time to save yourself time in the field.

SCALES USED FOR BUILDING PLANS

Inch Scale		Metric Scale
1/4	is closest to	1:50
1/8	is closest to	1:100

SCALE USED FOR SITE PLANS

Inch Scale		Metric Scale
1/16	is closest to	1:200

SYMBOLS FOR VARIOUS MATERIALS

Material	Chemical Symbol
Aluminum	AL
Antimony	Sb
Brass	—
Bronze	—
Chromium	Cr
Copper	Cu
Gold	Au
Iron (cast)	Fe
Iron (wrought)	Fe
Lead	Pb
Manganese	Mn
Mercury	Hg
Molybdenum	Mo
Monel	—
Platinum	Pt
Steel (mild)	Fe
Steel (stainless)	—
Tin	Sn
Titanium	Ti
Zinc	Zn

FIGURE 18.1   BIDET

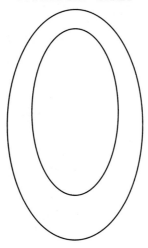

FIGURE 18.2   TOILET

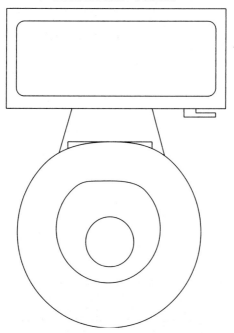

FIGURE 18.3   WHIRPOOL

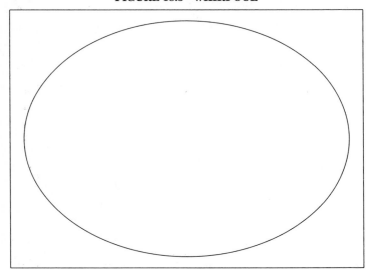

FIGURE 18.4   GARDEN TUB

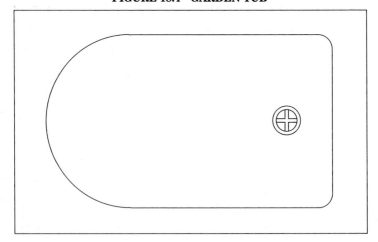

FIGURE 18.5   HOSE BIBB

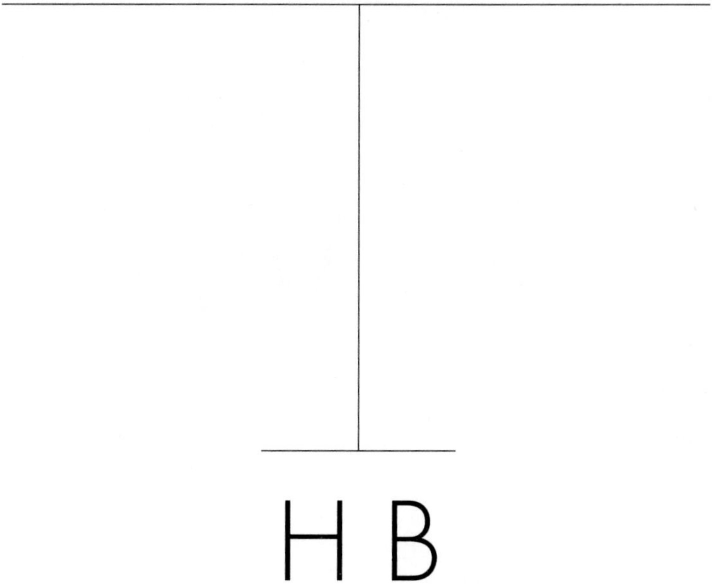

H B

FIGURE 18.6   FLOOR DRAIN

F D

FIGURE 18.7   WATER HEATER

FIGURE 18.8   SUMP PUMP

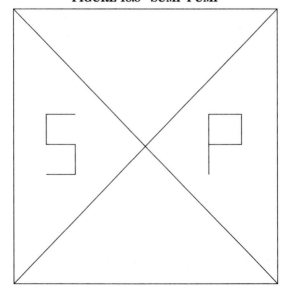

FIGURE 18.9   AUTOMATIC CLOTHES WASHER

FIGURE 18.10   DISHWASHER

FIGURE 18.11   CORNER SHOWER

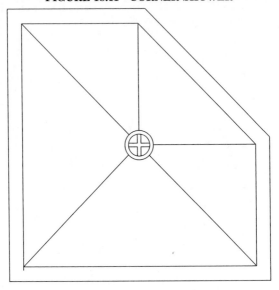

FIGURE 18.12   SHOWER

FIGURE 18.13   LAUNDRY TUB

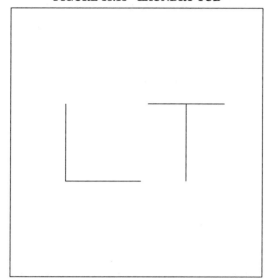

FIGURE 18.14   LAVATORY

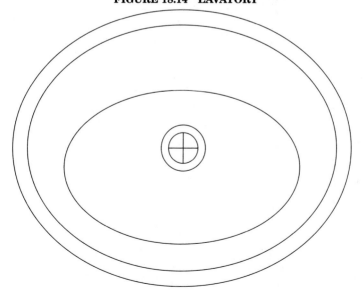

FIGURE 18.15 VANITY TOP

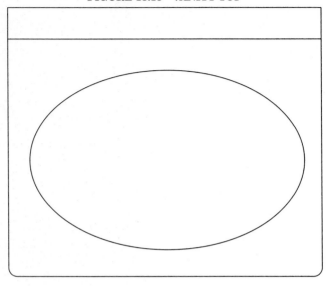

FIGURE 18.16 DOUBLE-BOWL VANITY TOP

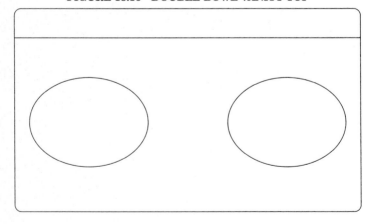

FIGURE 18.17 DOUBLE-BOWL KITCHEN SINK

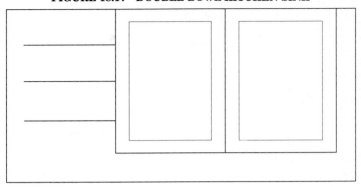

19

Safety

Safety is always a concern on construction jobs. This chapter provides you with some simple tips that may help you to avoid injury to yourself and others. The suggestions provided here are not inclusive of all safety factors to be considered. Depending upon the type of work you are doing and the type of job you are on, you may have to meet state and federal safety regulations. Failure to comply with mandated safety rules and regulations can result in injury, lawsuits, and fines. Don't take safety for granted. Invest some time in learning first aid, and always comply with safety requirements set forth on your jobs.

GENERAL SAFE WORKING HABITS

1. Wear safety equipment.
2. Observe all safety rules at the particular location.
3. Be aware of any potential dangers in the specific situation.
4. Keep tools in good condition.

SAFE DRESSING HABITS

1. Do not wear clothing that can be ignited easily.
2. Do not wear loose clothing, wide sleeves, ties, or jewelry (bracelets, necklaces) that can become caught in a tool or

otherwise interfere with work. This caution is especially important when working with electrical machinery.

3. Wear gloves to handle hot or cold pipes and fittings.

4. Wear heavy-duty boots. Avoid wearing sneakers on the job. Nails can easily penetrate sneakers and can cause a serious injury (especially if the nail is rusty).

5. Always tie shoelaces. Loose shoelaces can easily cause you to fall, possibly leading to injury to yourself or other workers.

6. Wear a hard hat on construction sites to protect the head from falling objects.

SAFE OPERATION OF GRINDERS

1. Read the operating instructions before starting to use the grinder.

2. Do not wear any loose clothing or jewelry.

3. Wear safety glasses or goggles.

4. Do not wear gloves while using the machine.

5. Shut the machine off promptly when you have finished using it.

SAFE USE OF HAND TOOLS

1. Use the right tool for the job.

2. Read any instructions that come with the tool unless you are thoroughly familiar with its use.

3. Wipe and clean all tools after each use. If any other cleaning is necessary, do it periodically.

4. Keep tools in good condition. Chisels should be kept sharp and any mushroomed heads kept ground smooth; saw blades should be kept sharp; pipe wrenches should be kept free of debris and the teeth kept clean.

5. Do not carry small tools in your pocket, especially when working on a ladder or scaffolding. If you should fall, the tools may penetrate your body and cause serious injury.

SAFE USE OF ELECTRIC TOOLS

1. Always use a three-prong plug with an electric tool.

2. Read all instructions concerning the use of the tool unless you are thoroughly familiar with its use.

3. Make sure that all electrical equipment is properly grounded. Ground fault circuit interrupters (GFCI) are required by OSHA regulations in many situations.

4. Use proper extension cords of a proper size. Undersized wires can burn out a motor, cause damage to the equipment, and present a hazardous situation.

5. Never run an extension cord through water or through any area where it can be cut, kinked, or run over by machinery.

6. Always hook up an extension cord to the equipment first and then plug it into the main electrical outlet, not vice versa.

7. Coil up and store extension cords in a dry area.

RULES FOR WORKING SAFELY IN DITCHES OR TRENCHES

1. Be careful of underground utilities when digging.

2. Do not allow people to stand on the top edge of a ditch while workers are in the ditch.

3. Shore all trenches deeper than 4 ft.

4. When digging a trench, be sure to throw the dirt away from the ditch walls.

5. Be careful to see that no water gets into a trench. Be especially careful in areas with a high water table. Water in a trench can easily undermine the trench walls and lead to a cave-in.

6. Never work in a trench alone.

7. Always have someone nearby, someone who can help you and locate additional help.

8. Always keep a ladder nearby so you can leave the trench quickly if need be.

9. Be watchful at all times. Be aware of any potentially dangerous situations. Remember, even heavy truck traffic nearby can cause a cave-in.

SAFETY ON ROLLING SCAFFOLDS

1. Do not lay tools or other materials on the floor of a scaffold. They can easily move and you could trip over them, or they might fall, hitting someone on the ground.

2. Do not move a scaffold while you are on it.

3. Always lock the wheels when the scaffold is positioned and you are using it.

4. Always keep the scaffold level to maintain a steady platform on which to work.

5. Take no shortcuts. Be watchful at all times, and be prepared for any emergencies.

WORKING SAFELY ON A LADDER

1. Use a solid and level footing to set up the ladder.

2. Use a ladder in good condition; do not use one that needs repair.

3. Be sure stepladders are opened fully and locked.

4. When using an extension ladder, place it at least 1/4 of its length away from the base of the building.

5. Tie an extension ladder to the building or other support to prevent it from falling or blowing down in high winds.

6. Extend a ladder at least 3 ft over the roof line.

7. Keep both hands free when climbing a ladder.

8. Do not carry tools in your pocket when climbing a ladder. (If you fall, the tools could cut into you and cause serious injury).

9. Use the ladder the way it should be used. For example, do not allow two people on a ladder designed for use by one person.

10. Keep the ladder and all its steps clean—free of grease, oil, mud, and so on—in order to avoid a fall and possible injury.

TO PREVENT FIRES

1. Always keep fire extinguishers handy, and be sure that the extinguisher is full and that you know how to use it quickly.

2. Be sure to disconnect and bleed all hoses and regulators used in welding, brazing, soldering, and the like.

3. Store cylinders of acetylene, propane, oxygen, and similar substances in an upright position in a well-vented area.

4. Operate all air-acetylene, welding, soldering, and related equipment according to the manufacturer's directions.

5. Do not use propane torches or other similar equipment near material that can easily catch fire.

6. Be careful at all times. Be prepared for the worst and be ready to act.

20

Electrical Motors

Plumbers sometimes have to work with electrical motors. This is often the case with plumbers who work in rural areas and with well pumps. Other motors, such as those on whirlpool tubs, can also require the attention of a plumber. You will find helpful information on this subject as you look through this chapter.

AC MOTORS

Motor Types	Split-Phase General Purpose	Split-Phase Special Service	Capacitor-Start Special Service	Capacitor-Start General Purpose	Polyphase 1 HP and Below
Starting torque (% full load torque)	130	175	250	350	275
Starting current	Normal	High	Normal	Normal	Normal
Service factor (% of rated load)	135	100	100	135	135

REMARKS

Split-phase general purpose	Low starting torque. High service factor permits continuous loading—up to 35 percent over nameplate rating. Ideal for applications of medium starting duty.
Split-phase special service	Moderate starting torque, but has service factor of 1. Apply where load will not exceed nameplate rating for any extended duration of time. Because of higher starting current, use where starting is infrequent.
Capacitor-start special service	High starting torque but has 1 service factor. Use only where load will not exceed nameplate rating for any extended duration of time. Starting current is normal.
Capacitor-start general purpose	Very high starting torque. High service factor permits continuous loading up to 35 percent over nameplate rating. Ideal for powering devices with heavy loads, such as conveyors.
Polyphase 1 HP and below	Normal start current for polyphase is low compared to single-phase motors. High starting ability. High service factor permits continuous loading up to 35 percent over nameplate rating. Direct companion to general purpose capacitor-start motor.

MOTOR TROUBLESHOOTING GUIDE

Symptom	Possible Causes	Correction
High input current (all three phases)	Accuracy of ammeter readings.	First check accuracy of ammeter readings on all three phases.
Running idle (Disconnected from load)	High line voltage. 5 to 10 percent over nameplate.	Consult power company. Possibly decrease by using lower transformer tap.
Running loaded	a. Motor overload. b. Motor voltage rating does not match power system voltage.	a. Reduce load or use larger motor b. Replace motor with one of correct voltage rating. c. Consult power company. Possibly correct by using a different transformer tap.

MOTOR TROUBLESHOOTING GUIDE (cont.)

Symptom	Possible Causes	Correction
Unbalanced input current (5 percent or more deviation from the average input current)[a]	Unbalanced line voltage due to: a. Power supply. b. Unbalanced system loading. c. High resistance connection. d. Undersized supply lines.	Carefully check voltage across each phase at the motor terminals with a good, properly calibrated voltmeter.
	Defective motor	If there is doubt as to whether the trouble lies with the power supply or the motor, check as follows: Rotate all three input power lines to the motor by one position—i.e., move line #1 to #2 motor lead, line #2 to #3 motor lead, and line #3 to #1 motor lead. a. If the unbalanced current pattern follows the input power lines, the problem is in the power supply. b. If the unbalanced current pattern follows the motor leads, the problem is in the motor. Correct the voltage balance of the power supply or replace the motor, depending on answer to a and b above.

[a] A small voltage imbalance can produce a large current imbalance. Depending on the magnitude of imbalance and the size of the load, the input current in one or more of the motor input lines may greatly exceed the current rating of the motor.

MOTOR TROUBLESHOOTING GUIDE (*cont.*)

Symptom	Possible Causes	Correction
Excessive voltage drop	Excessive starting or running load.	Reduce load.
	Inadequate power supply.	Consult power company.
	Undersized supply lines.	Increase line sizes.
	High resistance connections.	Check motor leads and eliminate poor connections.
	Each phase lead run in separate conduits.	All 3-phase leads must be in a single conduit, per National Electrical Code. (This applies only to metal conduit with magnetic properties.)
Overload relays tripping upon starting	Slow starting (10–15 sec or more) due to high inertia load.	Reduce starting load. Increase motor size if necessary.
	Low voltage at motor terminals.	Improve power supply and/or increase line size.

MOTOR TROUBLESHOOTING GUIDE (cont.)

Symptom	Possible Causes	Correction
Running loaded	Overload.	Reduce load or increase motor size.
	Unbalanced input current.	Balance supply voltage.
	Single phasing.	Eliminate.
	Excessive voltage drop.	Eliminate (see above entry).
	Too frequent starting or intermittent overloading.	Reduce frequency of starts and overloading or increase motor size.
	High ambient starter temperatures.	Reduce ambient temperature or provide outside source of cooler air.
	Wrong size relays.	Correct size per nameplate current of motor. Relays have built-in allowances for service factor current. Refer to National Electrical Code.
Motor runs excessively hot	Overloaded.	Reduce load or load peaks and number of starts in cycle or increase motor size.
	Blocked ventilation.	Clean external ventilation system. Check fan. Blow out internal ventilation passages. Eliminate external interference to motor ventilation.
	High ambient temperature over 40°C (104°F).	Reduce ambient temperature or provide outside source of cooler air.
	Unbalanced input current.	Balance supply voltage. Check motor leads for tightness.
	Single phased.	Eliminate single phase condition.

MOTOR TROUBLESHOOTING GUIDE (cont.)

Symptom	Possible Causes	Correction
Won't start (just hums and heats up)	Single phased.	Shut power off. Eliminate single phasing. Check motor leads for tightness.
	Rotor or bearings locked.	Shut power off. Check shaft for free-ness of rotation.
		Be sure proper sized overload relays are in each of the 3 phases of starter. Refer to National Electrical Code.
Runs noisily under load	Single phased.	Shut power off. If motor cannot be restarted, it is single phased. Eliminate single phasing.
		Be sure proper sized overloaded relays are in each of the 3 phases of the starter. Refer to National Electrical Code.
	Overload.	Reduce load or increase voltage.
Load speed appreciably below nameplate speed	Excessively low voltage.	*Note:* A reasonable overload or voltage drop of 10–15 percent will reduce speed only 1–2 percent.
		A report of any greater drop would be questionable.
	Inaccurate method of measuring RPM.	Check meter using another device or method.

MOTOR TROUBLESHOOTING GUIDE (*cont.*)

Symptom	Possible Causes	Correction
Excessive vibration (mechanical)	Out of balance:	
	a. Motor mounting.	Be sure motor mounting is tight and solid.
	b. Load.	Disconnect belt of coupling. Restart motor. If vibration stops, the unbalance was in load.
	c. Sheaves or coupling.	Remove sheave or coupling. Securely tape 1/2 key in shaft keyway and restart motor. If vibration stops, the unbalance was in the sheave or coupling.
	d. Motor.	If the vibration does not stop after checking a, b, and c above, the unbalance is in the motor. Replace the motor.
	e. Misalignment on close coupled application.	Check and realign motor to the driven machine.
Noisy bearings (listen to bearings)		
Smooth mid-range hum	Normal fit.	Bearing OK.
High whine	Internal fit of bearing too tight.	Replace bearing. Check fit.
Low rumble	Internal fit of bearing too loose.	Replace bearing. Check fit.

MOTOR TROUBLESHOOTING GUIDE (*cont.*)

Symptom	Possible Causes	Correction
Rough clatter	Bearing destroyed.	Replace bearing. Avoid the following: a. Mechanical damage. b. Excessive greasing. c. Wrong grease. d. Solid contaminants. e. Water running into motor. f. Misalignment or close coupled application. g. Excessive belt tension.
Mechanical noise	Driven machine or motor noise.	Isolate motor from driven machine. Check difference in noise level.
	Motor noise amplified by resonant mounting.	Cushion motor mounting or dampen source of resonance.
	Driven machine noise transmitted to motor through drive.	Reduce noise of driven machine or dampen transmission to motor.
	Misalignment or close coupled application	Improve alignment.

21

Definitions

Consider this chapter as your dictionary of plumbing words, terms, and phrases. Many of the entries will be familiar to experienced plumbers, but I suspect there are some words that you might not be able to identify on your own. Take a little test. Look at the lead words and see if you know what they all mean. If you're involved in the trade, it will pay to know the language used in and around it.

Abrasive Any material that erodes another material by rubbing.

ABS (Acrylonitrile Butadiene Styrene) A plastic material used for drainage.

Adapter A fitting that joins pipes of different materials of different sizes.

Aerator A device that adds air to water; it fills flowing water with bubbles to avoid splashing.

Air chamber A device designed to absorb the shock of a fast-closing valve or faucet.

Air gap The unobstructed vertical distance through the air between the lowest outlet from any pipe or faucet supplying water to a tank, plumbing fixture, or other device and the flood level rim of the receptacle.

Anaerobic bacteria   Bacteria that exist in the absence of free oxygen (air).

Anchor   A special fastener used to attach pipes, fixtures, and other parts to the building structure.

Angle valve   A globe valve in which the inlet and outlet openings are at 90° angles to each other.

Apprentice plumber   A person who is learning the plumbing trade by practical experience.

Area drain   A drain installed to receive surface or rain water from an open area.

Asbestos joint runner   A runner made of an asbestos rope and a clamp that holds molten lead in the bell of a cast pipe until it has cooled.

Back fill   Material used to fill an excavated trench.

Back-flow   The flow of water in pipes in a reverse direction from that normally intended.

Back-flow connection   Any connection or arrangement by which back-flow may occur.

Back-flow preventer   A device that prevents back-flow into the potable water supply system.

Backing   Wood or other supports placed in the building walls to which plumbing fixtures and other equipment can be attached.

Back vent   A branch vent connected to the main vent stack and extending to a location near a fixture trap.

Backwater valve   A type of check valve installed to prevent the back-flow of sewage from flooding the basement or lower levels of a building or dwelling.

Ballcock   A valve or faucet controlled by a change in the water level; it is primarily associated with toilet tank operation.

Ball valves   A valve in which the flow of fluid is controlled by a rotating drilled ball that fits tightly against a resilient (flexible) seat in the valve body.

Basket strainer   A kitchen sink drain fitting, also called a duo-strainer.

Battery of fixtures   Any group of two or more similar adjacent fixtures.

Bearing partition   An interior wall of a building that carries the load of the structure above in addition to its own weight.

Bell or hub   The enlarged end of some types of cast-iron pipe that fits over the next pipe section.

Benchmark   A fixed location of known elevation.

Bend   A change of direction in piping.

Bib   Another name for faucet.

Bidet   A bowl equipment with cold and hot running water used for bathing the external genitals and posterior parts of the body.

Blowoff   The controlled discharge of excess pressure and temperature.

Blueprint   Drawings with accurate measurements that are used to install piping and building materials.

BOCA   Building Officials Conference of America.

Bonnet   The upper portion of the gate valve body.

Bracket hanger   A hanger supporting a wall-hung sink or fixture.

Branch   An addition to the main pipe in a piping system.

Branch vent   A vent that connects a branch of the drainage piping to the main vent stack.

Braze   A means of joining metal with an alloy having a melting point higher than common solder but lower than the metal being brazed.

Building drain   That part of the lowest piping of the drainage system that receives the discharge from soil, waste, and other draining pipes inside the walls of the building and conveys it to the building sewer.

Building drainage system   The complete system of pipes installed for the purpose of carrying waste water and sewage to septic or sanitary sewer systems.

Building drain branch   A soil or waste pipe that extends horizontally from the building drain and receives only the discharge from fixtures on the same floor as the branch.

Building main   Water supply piping that begins at the property line and ends in the building itself.

Building sanitary drain   A building drain that conveys sewage only.

Building sewer   That part of the drainage system that extends from the end of the building drain and conveys its discharge to the public sewer system.

Building storm drain   A building drain that conveys storm water only.

Building storm sewer   A building sewer that conveys storm water but not sewage.

Building trap   A trap placed in the building drain to prevent entry of sewer gases from the sewer main.

Burr   A sharp, rough edge on a piece of pipe or tubing as a result of being cut.

Bushing   A pipe fitting with both male and female threads; it is used to connect pipes of different sizes.

Cap   A female pipe fitting that is closed at one end. It is used to close off the end of a piece of pipe or tubing.

Capillary attraction   The movement of liquid upward.

Cast-iron pipe   Any pipe made from cast iron.

Caulk   The material used to seal joints.

Caulking   A method of making a bell and spigot pipe joint watertight by packing it with oakum and lead.

Cesspool (dry well)   A deep pit that receives liquid waste and permits the excess liquid to be absorbed into the ground.

Chain wrench   An adjustable tool for holding and turning large pipe up to 4 in. in diameter. A flexible chain replaces the steel jaws of two standard pipe wrenches.

Chalk line   A marking tool consisting of a string coated with chalk.

Change in direction   The term applied to the various turns that may be required in drainage pipes and other piping systems.

Chase (a pipe chase)   A space or recess in the walls of a building where pipes are run.

Check valve   A device preventing back-flow in pipes. Water can flow readily in one direction but any reversal of the flow causes the check valve to close.

Circuit vent   A branch vent that functions for two or more traps.

Clean-out   A fitting with a removable plug that is placed in drainage pipe to allow entry into the system in order to relieve stoppages.

Close nipples   The shortest length of a given size pipe that is threaded on both ends.

Closet bend   An elbow drainage fitting connecting a water closet to the drainage system.

Closet bolt   A bolt used to attach a water closet securely to the closet flange.

Closet spud   The connector between the base of the ballcock assembly in a water closet tank and the water supply pipe.

Code   A set of regulations that has been adopted by a governmental unit for the purpose of protecting the public health and safety. In plumbing, these codes regulate the quality of materials, the design and installation of plumbing systems, and the method used to test the systems.

Cold chisel   A handtool used with a hammer to cut cast iron or concrete.

Common vent   A vent at the junction of two fixture drains that serves as a single vent for both fixtures.

Compression faucet or valve   A faucet or valve designed to stop the flow of water by the action of a flat disk (washer) closing against a seat.

Continuous vent   A vertical vent that is a continuation of the drain to which it connects.

Continuous waste   The waste from two or three fixtures connected to a single trap.

Copper pipe straps   Straps made from copper that are used to secure copper pipe.

Corporation stop   A valve installed in the building water service line at the water main; it is also called corporation cock.

Counterflashing   A flashing usually used on chimneys to prevent entry of moisture.

Coupling   A pipe fitting containing female threads on both ends. Couplings are used to join two or more lengths of pipe in a straight run.

CPVC (chlorinated polyvinyl choride)   A type of plastic used to make pipe that will carry hot water, air, or chemicals.

Crawl space   The space between the floor framing and the ground in a building that has no basement.

Cross   A pipe fitting with four female openings at right angles to one another.

Cross connection   Any link between contaminated water and potable water in the supply system.

Crossover   The connection of two piping runs in the same piping system or the connection of two different piping systems that contain potable water.

Crown of a trap   The point in a trap where the direction of flow changes from upward to downward.

Crown weir   The point in the curve of the trap directly below the crown.

Curb box   A cylindrical casting placed in the ground over the corporation stop. It extends to ground level and permits a special key to be inserted to turn off the corporation cock.

Curb cock or curb stop   A valve placed on the water service, usually near the curb line.

Dead end   A branch of a drainage piping system that ends in a closed fitting.

Deep seal trap   A trap located in the building drain to re-sist back-pressure of sewer gas.

Developed length   The length of pipe and fittings mea-sured along the center line.

Die   A tool used to cut external threads by hand or ma-chine.

Die stock   A tool used to turn dies when cutting external threads.

Dip of a trap   The lowest portion of the inside top surface of the trap.

Dope   A pipe joint compound.

Double hub   A cast-iron sewer pipe having a bell on both ends.

Downspout   A vertical pipe made from sheet metal, cop-per or plastic that carries water from the gutters to the ground or to a storm drain.

Drain   Any pipe that carries wastewater or waterborne wastes.

Drain, building   Horizontal piping that connects the building drainage piping to the sanitary sewer or private sewage system.

Drainage fitting   A pipe fitting designed for use with drainage piping.

Drainage fixture unit (dfu)   A measure of the probable discharge into the drainage system by various types of plumbing fixtures on the basis of one dfu being equal to 7.5 gpm discharge. The drainage fixture-unit valve for a particular fixture depends on its volume rate of drainage discharge.

Drainage piping   All or any portion of the drainage pip-ing system.

Drainage system   The piping within public or private premises that conveys sewage or other liquid waste to a legal point of connection to a public sewer system or pri-vate disposal system.

Drains, storm   Piping systems that carry water and rain-water from a building to the storm sewer.

Drinking fountain   A fixture that delivers a stream or jet of drinking water through a nozzle.

Drum trap   A trap whose main body is a cylinder with its axis vertical. This cylinder is larger in diameter than the inlet or outlet pipe.

Dry vent   Any vent that does not carry waste water.

Dry well   A well fitted with aggregate, designed to permit water to seep into the ground; used to receive rain water.

Ductility   The property of a material that allows it to be formed into thin sections without breaking.

Easement   The right to use land owned by another for some specific purpose (for example, the right of a public utility or municipality to install service through a person's property).

Eccentric fitting   A pipe fitting in which the centerline of the openings is offset.

Effluent   The outflow from sewage treatment equipment.

Eighth bend   A pipe fitting that causes the run of pipe to make a 45° turn.

Elbow   A pipe fitting having two openings, which causes a run of pipe to change directions 90°.

Erosion   The gradual wearing away of material as a result of abrasive action.

Evaporation   Loss of water to the atmosphere.

Excavation lines   Lines laid out on the job site to indicate where digging for foundation and piping is to be done.

Existing work   That part of the plumbing system that is in place when an addition or alteration is begun.

Expansion joint   A joint that permits pipe to move as a result of expansion.

Extra heavy   A term used to designate the heaviest and strongest grades of cast-iron and steel pipe.

Fall   The amount of slope given to horizontal runs of pipe.

Faucet   A valve the purpose of which is to permit a controlled amount of water from the water pipe.

Female thread   Any internal thread.

Ferrule   A cast-iron fitting installed in the bell of a cast-iron pipe.

Field tile   Short lengths of clay pipe that are installed as subsurface drains.

Fill   Sand, gravel, or other loose earth.

Finishing   The third major stage of the plumbing process.

Fittings   The parts of the piping system that serve to join lengths of pipe.

Fixture   A device such as a sink, lavatory, bathtub, water closet, or shower stall.

Fixture branch   The water supply piping that connects a fixture to the water supply piping.

Fixture drain   Drainage piping, including a trap that connects a fixture and a branch waste pipe.

Fixtures, battery   Any two or more similar fixtures served by the same horizontal run of drainage piping.

Fixtures, combination   A fixture, such as a kitchen sink/laundry basin that is specifically designed to perform two or more functions.

Fixture supply pipe   The water supply pipe that connects the fixture to the stub-out.

Fixture unit   A means of rating the amount of discharge from a given fixture so that the drainage piping is large enough to carry the required amount of waste. (Also: a flow of 1 cfm.)

Fixture vent   A part of the piping system that connects with the drainage piping near the point where the fixture trap is installed and extends to a point above the roof of the structure.

Flange   A rim of collar attached to one end of a pipe to give support or a finished appearance.

Flange nut   A device that connects flared copper pipe to a threaded flare fitting.

Flashing   Materials such as copper or stainless steel that are installed as joints between roofs and walls and roofs and chimneys to prevent water from entering the structure.

Float arm   A thin rod threaded at each end that connects the float ball to the inlet valve of the ballcock assembly in a toilet tank.

Float ball   A metal or plastic ball used to control the inlet valve in water closet tanks.

Flooded   A condition that occurs when liquid rises to the flood level of a fixture.

Flood level   The point in a fixture above which water overflows.

Floor drain   A fitting located in the floor to carry waste water into the drainage piping.

Floor flange   A fitting attached at floor level to the end of a closet bend so that the water closet can be bolted to the drainage piping.

Flow pressure   The pressure in the water supply pipe near the faucet or water outlet while the faucet or water outlet is wide open and flowing.

Flow rate   The volume of water used by a plumbing fixture in a given amount of time. Usually expressed in gallons per minute (gpm).

Flush   To clean by drenching with a large amount of water.

Flush ball   In a water closet tank assembly, the rubber ball-shaped object that controls the flow of water into the bowl.

Flush bushing   A pipe fitting used to reduce the diameter of a female-threaded pipe fitting.

Flushometer   A valve that permits a preestablished amount of water to enter a fixture such as a water closet or urinal.

Flush tank   A receptacle designed to discharge, either manually or automatically, a predetermined quantity of water to fixtures for flushing purposes.

Flush valve   A device for flushing water closets and similar fixtures.

Flush valve seat   The opening between the tank and bowl in a water closet against which the flush ball is fitted.

Flux   A chemical substance that prevents oxides from forming on the surface of metals as they are heated for soldering, brazing, and welding.

Footing   The part of the foundation of a building that rests directly on the ground. The footing distributes the weight of the building over a sufficiently large amount of ground so that the building will not settle excessively.

Force cup   A rubber cup attached to a wooden handle; it is used for unclogging water closets and drains. It is also called a plunger or "plumber's friend."

Foundation   That part of a building which is below the first framed floor and includes the foundation wall and footing.

Foundation drain   Piping around the base of the foundation to collect ground water and convey it into a sump.

Freezeless water faucet   A water faucet designed to be installed through an exterior wall to prevent freezing.

Front main clean-out   A plugged fitting located near the front wall of a building where the building drain leaves the building. The front main clean-out may be either inside or directly outside the building foundation wall.

Frostline   The depth of frost penetration in the soil. This depth varies in different parts of the country, depending on the normal temperature range.

Frost-proof closet   A closet that has no water in the bowl and has the trap and flush valve installed below ground level, usually below the frost line.

Full bath   A bathroom containing a water closet, a lavatory, and a bathtub.

Galvanized iron   Iron that has been coated with zinc to prevent rust.

Garbage disposal   An electric grinding device used with water to grind food wastes into pulp and discharge the pulp into the drainage system.

Gasket   Any semihard material placed between two surfaces to make a watertight seal when the surfaces are drawn together by bolts or other fasteners.

Gate valve   A valve that uses a disc moving at a right angle to the flow of water to regulate the rate of flow. When a gate valve is fully opened, there is no obstruction to the flow of water.

Globe valve   A spherically shaped valve body that controls the flow of water with a compression disc. The disc, opened and closed by means of a stem, mates with a ground seat to stop water flow.

Grade or pitch   The fall (slope) of a line of pipe in reference to a horizontal plane. As applied to plumbing drainage, pitch is usually expressed as the fall in a fraction of an inch per foot length of pipe (for example: 1/4 in./ft).

Grease interceptor   A receptacle designed to separate and retain grease and fatty substances from wastes normally discharged from kitchens.

Ground water   Water in the subsoil.

Hacksaw   A metal-cutting saw with a replaceable blade.

Half bath   A bathroom containing a water closet and a lavatory.

Handle puller   A tool for removing handles from faucets and valves.

Hanger   A support for pipe.

Header   A water supply pipe to which two or more branch pipes are connected to service fixtures.

Headroom   Space between the floor and the lowest pipe, duct, or part of the framing.

Horizontal branch   Any horizontal pipe in the waste piping system that extends from a stack to the fixture trap.

Horizontal pipe Any pipe that is installed so that it makes an angle of less than 45° from level.

Hose bib A water faucet made with a threaded outlet for the attachment of a hose.

House drain The horizontal part of the drainage piping that connects the piping system within the structure to the sanitary sewer or private sewage treatment equipment.

Hub The enlarged end of a hub-and-spigot cast-iron pipe.

Hydrant Water supply outlet with a valve located below ground.

Increaser A fitting installed in a vent stack before the stack goes through the roof. It enlarges the stack or vent pipe.

Indirect waste pipe A waste pipe that does not connect directly with the drainage system but discharges into it through a properly trapped fixture or receptacle.

Individual vent A pipe installed to vent a fixture trap that connects with the vent system above the fixture it serves.

Industrial waste A liquid waste resulting from the processes employed in industrial establishments.

Interceptor A receptacle designed and constructed to intercept, separate, and prevent the passage of detrimental floating or heavy solids.

Invert The lowest portion of the inside of any horizontal pipe.

Joint runner A tool composed of asbestos rope and a clamp used in leading joints in horizontal runs of bell-and-spigot cast-iron pipe.

Journeyman plumber A person who has acquired the requisite skill and knowledge necessary for the proper installation of plumbing. The requirements for this title are four years of training and experience under the supervision of a licensed master plumber, or the equivalent thereof in education, training, and experience.

K grade copper tube   Copper tubing suitable for installation underground.

Laundry tray   A fixed tub, installed in a laundry room of a house, that is supplied with cold and hot water and a drain connection and is used for washing clothes and other household items.

Lavatory   A fixture designed for washing hands and face.

Lay out   The act of measuring and marking the location of something.

Layout   The arrangement of a house, room, or part of a job.

Leach bed   A system of underground piping that permits absorption of liquid waste into the earth; also called disposal field or leach field.

Leader   A pipe from a roof drain to a storm drain of a building.

Level   A tool used to determine if something is horizontal or vertical.

L grade copper tube   A type of copper tube that may be used to supply potable water.

Line level   A small, lightweight level designed to be hung from a string line to determine if the line is horizontal.

Liquid waste   The liquid discharge from a plumbing fixture.

Long quarter bend   A 90° fitting with one section longer than the other.

Long-sweep fitting   Any drainage fitting that has a long radius curve at the bends.

Loop vent   A branch vent similar to a circuit vent except that it connects with the stack vent instead of the vent stack.

Lot line   The line(s) forming the legal boundary of a piece of property.

Main   The principal pipe artery to which branches may be connected.

Main sewer   The large sewer to which the building drains of several houses are connected.

Main vent   The principal artery of the venting system to which vent branches may be connected.

Main water line   The large water supply pipe to which branches are connected.

Male thread   Threads on the outside of a pipe, fitting, or valve.

Malleable iron   Cast iron that has been heat treated to reduce its brittleness.

Mallet   A soft-face hammer (rawhide or plastic) used to drive parts without damaging them.

Manhole   An opening in the sanitary or storm sewer system to permit access.

Masonry bit   A bit designed to drill holes in mortar, tile, and concrete.

Master plumber   A person who has had at least two years of experience as a journeyman plumber and is licensed as a master plumber and is engaged in the business of plumbing.

Meter stop   A valve used on a water main between the street and a water meter.

Miter box   A hardwood or metal saw guide. The sides are slotted to guide a hand saw for 45° and 90° cuts.

Mixing faucet   Separate faucets having a common spout permitting control of the water temperature.

Moisture barrier   A material such as polyethylene that retards the passage of vapor or moisture into walls or through concrete floors.

Mop basin   A floor set service sink; also called a mop receptor.

Negative pressure   A pressure within a pipe that is less than atmospheric pressure.

Neoprene   A synthetic rubber with superior resistance to oils; often used as gasket and washer material.

Nipples Short lengths of pipe (usually less than 12 in.) with male threads on both ends that are used to joint fittings.

No-hub pipe Soil pipe that has smooth ends but doesn't have a spigot or hub.

Nominal size The approximate dimension(s) of standard material.

Nonbearing wall A wall within a structure that supports no load other than its own weight.

Nonrising stem valve A gate valve in which the stem does not rise when the valve is opened.

Nozzle A fitting attached to the outlet of a pipe or hose that varies the volume of water and causes the shape of the stream of water to be changed to a spray of varying diameter.

Oakum Loosely woven hemp rope that has been treated with oil or waterproofing agent; it is used to caulk joints in a bell-and-spigot pipe and fitting.

Offset A combination of elbows or bends that permits a section of a pipe to be out of line but in a line parallel with its original alignment.

O-ring A rubber seal used around stems of some valves to prevent water from leaking past.

Outside wall Any wall of a structure exposed to the weather on one side.

Overflow tube A vertical tube in a water closet tank that prevents overfilling of the tank.

Oxidized sewage Sewage that has been exposed to oxygen to make the organic substances stable.

Packing A loosely packed waterproof material installed in the packing box of valves to prevent leaking around the stem.

Packing nut A special nut holding the stem in a faucet or valve while compressing the packing.

Partition or partition wall An interior wall that divides spaces within a building.

Petcock   A small ground key type valve used with soft copper tubing.

Pilot light   A relatively small flame that burns constantly. Its purpose is to ignite the main supply of gas.

Pipe   A cylindrical conduit or conductor, the wall thickness of which is sufficient to receive a standard pipe thread.

Pipe die   A tool for cutting external pipe threads.

Pipe joint compound   Material used for sealing threaded pipe joints.

Pipe, soil   A pipe for conveying waste that contains fecal matter (human waste).

Pipe strap   A metal strap used to support or hold pipe in place.

Pipe, vertical   Any pipe or part thereof that is installed in a vertical position.

Pipe, waste   A pipe that conveys only liquid and other waste, not fecal matter.

Pipe, water distribution   Pipes that carry water from the service pipe to fixtures in the building.

Pipe, water riser   A water supply pipe that rises vertically from a horizontal pipe.

Pipe wrench   A wrench with adjustable, slightly curved, toothed jaws designed to grip pipe firmly as pressure is applied to the handle.

Pipes, water service   That portion of the water piping that extends from the main to the meter.

Piping   A generic term used to refer to all the pipes in a building.

Pitch   The degree of slope or grade given a horizontal run of pipe.

Plug   A pipe fitting with external threads and a head that is used for closing the opening in another fitting.

Plumb   Exactly perpendicular (vertical); at a right angle to the horizontal.

Plumb bob   A tool consisting of a weight suspended by a string. When allowed to hang freely, the string line assumes a position that is exactly vertical.

Plumber   A person trained and experienced in the skill of plumbing.

Plumber's friend   A plunger, or force cup; a tool consisting of a rubber cup and handle used underwater to force a blockage through sewer lines.

Plumber's furnace   A heating source used to melt lead, heat soldering irons, or solder.

Plumbing   The art of installing in buildings the pipes, fixtures, and other apparatus for bringing in the water supply and removing waste water and water-carried waste.

Plumbing appliance   A special class of plumbing fixture intended to perform a special function.

Plumbing fixture   A receptacle for wastes that are ultimately discharged into the sanitary drain system.

Plumbing inspector   A person authorized to inspect plumbing and drainage for compliance with the code for the municipality.

Plumbing system   The plumbing system of a building—including the water supply distributing pipes; the fixtures and fixture traps; the soil, waste, and vent pipes; the building drain and building sewer; and the storm water drainage—with its devices, appurtenances, and connections within the outside the building within the property line.

Polyethylene   A plastic used to make pipe and fittings for underground water systems.

Pool   A permanently installed water receptacle used for swimming, diving, or bathing, designed to accommodate more than one person at a time.

Pop-off valve   A safety valve that opens automatically when pressure and temperature exceed a predetermined limit.

Porcelain   A white ceramic material used for bathroom fixtures; it is also called vitreous enamel.

Port control faucet   A single-handle, noncompression faucet that contains within the faucet body a port for both cold and hot water and some method of opening and closing these ports.

Positive pressure A pressure within the sanitary drainage or vent piping system that is greater than atmospheric pressure.

Potable water Water from a public water supply approved by a state department of health or a private water supply that has been accepted by an administrative authority as satisfactory for human consumption.

Potable water supply system The water service pipe, the water distributing pipes, and the necessary connecting pipes, fittings, control valves, and all appurtenances within the building or outside the building that are within the property lines.

Precipitation The total measurable amount of water received in the form of snow, rain, hail, and sleet. It is usually expressed in inches per day, month, or year.

Pressure head The amount of force or pressure created by a depth of 1 ft of water.

Pressure regulator A valve that reduces water pressure in the supply piping.

Private sewer A sewer system privately owned and not directly under the jurisdiction of a municipality or a public utility.

Propane Hydrocarbon derived from crude petroleum and natural gas and used as a fuel for plumber's furnaces or torches.

P-trap A trap commonly used on plumbing fixtures.

Public potable water supply A water supply approved by a state department of health.

Pubic sewer A sewer system approved by a state department of health and located in a street, alley, or other premises under the jurisdiction of a municipality or a public utility.

Punch list A list, made by the builder or owner near the end of construction, indicating what must be done before a house is completely finished and ready for occupancy.

Putty A soft, prepared mixture used to seal sink rims, water closet bases, and other places where a sealant is needed.

PVC (polyvinyl chloride) A type of plastic used to make plumbing pipe and fittings for water distribution, irrigation, and natural gas distribution.

Quarter bend A drainage pipe fitting that makes a 90° angle.

Rainwater leader A pipe inside the building that conveys storm water from the roof to a storm drain; also called a conductor or downspout.

Reamer A tool used in reaming.

Reaming Removing the burr from the inside of a pipe that has been cut.

Recovery rate Speed at which a water heater will heat cold water to the desired temperature.

Reducer A pipe fitting having one opening smaller than the other. Reducers are used to change from a relatively large diameter pipe to a smaller one.

Refill tube A copper or rubber tube extending from the ballcock to the overflow tube in the water closet assembly.

Reinforcement wire Heavy woven wire placed in concrete to give added strength.

Reinforcing rod Embossed steel rods placed in concrete slabs, beams, or columns to increase their strength.

Relief valve A safety device that automatically provides protection against excessive temperatures, excessive pressures, or both.

Relief vent A branch from a vent stack, connected to a horizontal branch between the first fixture drain and a soil or waste stack.

Return offset An offset that permits a pipe to be returned to its original alignment.

Rigid copper tubing Hard copper used when installing water lines.

Rim The unobstructed open edge of the receptacle section of a plumbing fixture.

Riser The water supply pipe that extends vertically for the height of one full story or more and from which water is supplied to fixture branches.

Rising stem   A type of valve stem that moves up and down as the valve is opened and closed.

Roof drain   A drain installed in a flat or nearly flat roof to receive water and conduct it into a leader, downspout, or conductor.

Roof jacket or flange   A jacket or flange installed on the roof terminals of vent stacks and stack vents to seal the opening and prevent rainwater from entering into the building around the vent pipe.

Rotating ball faucet   A single-handed faucet that controls water flow and temperature with a channeled rotating plastic ball. Holes in the ball are aligned with orifices for hot and cold water.

Rough-in   Earliest stage of plumbing installation, sometimes divided into two stages: first rough brings water and sewer lines inside the building foundation; second rough is the installation of all piping that will be enclosed in the walls of the finished building.

Rough-in measurements   Measurements that indicate where the water supply and waste piping must terminate in order to serve the fixtures that will be installed later.

Run   One or more lengths of pipe that continue in a straight line.

Running trap   A stretch of pipe in which the inlet and outlet are at the same height and the waterway between them is lower than the bottom of either.

Saddle fitting   A fitting used to install a branch from an existing run of pipe.

Safety valve   A combination temperature and pressure relief valve generally installed in a hot-water tank to prevent an explosion caused by overheating or excessive pressure inside the tank.

Sand trap or interception   A device designed to allow sand and other heavy particles to settle out before water enters the water supply piping.

Sanitary drainage pipe   Pipes installed to remove waste water and waterborne wastes.

Sanitary sewage Water and waterborne waste containing human excrement as well as other liquid household wastes.

Sanitary T-branch A drainage fitting having three openings and formed in the shape of a T.

Sanitary Y-branch A drainage fitting shaped like a Y.

Scaffold Any platform erected temporarily to support workers and materials while work is being done.

Scale drawing A drawing of any object that has been carefully reduced to a fraction of the real size so that all parts are in the correct proportion.

Scuttle A small opening in a ceiling that provides access to an attic or roof.

Seal of a trap The depth of water held in a trap under normal operating conditions.

Secondary branch Any branch off the primary branch of a building drain.

Self-syphonage The loss of the seal of a trap as a result of removing the water from the trap. It is caused by the discharge of the fixture to which the trap is connected.

Septic tank A watertight tank in a private waste disposal system that receives household sewage. Within the septic tank, solid matter is separated from the water before the water is discharged.

Service L (Street L) A 45° or 90° elbow with external threads on one end and internal threads on the other.

Service pipe The water supply pipe from the main in the street or other source of supply to the building.

Service sink A sink with a deep basin to accommodate a scrub pail. It is used for the filling and emptying of scrub pails, the rinsing of mops, and the disposal of cleaning water; it is also called a slop sink.

Sewage All water and waterborne waste discharged through the fixture.

Sewerage A piping system designed to convey sewage.

Sewer, building (house sewer) Horizontal sewage piping that extends from the building to the sewer main.

Sewer, building storm   The piping from the building storm drain to the public storm sewer.

Sewer, private   A sewer owned and maintained privately. It may contain sewage from building(s) to a public sewer or to a privately owned sewage disposal system.

Sewer, storm   A sewer used to carry rainwater, surface water, or similar water wastes that do not include sanitary sewage.

Sewer gas   The mixture of vapors, odors, and gases found in sewers.

Shut-off valve   A valve installed in a waterline whenever a cut-off is required.

Side outlet   An opening at the side of a fitting; a T or Y fitting having one side opening.

Side vent   A vent connected to a drain at an angle of 45° or less.

Sillcock   A faucet used on the outside of a building to which a garden hose can be attached.

Single lever faucet   Any of several types of washerless faucets using a single control and springs, balls, or cartridges to control the flow and temperature of the water.

Siphonage   A partial vacuum crated by the flow of liquids in pipes.

Size of pipe   The nominal dimension by which the pipe is designated; approximately equal to the inside diameter of the pipe.

Slab   A large, flat, concrete section such as a basement floor, driveway, or patio.

Slip coupling   A pipe coupling that has no stop to prevent it from slipping over a pipe.

Slip joint   A connection in which one pipe slides inside another.

Slip nut   A nut used on P traps and similar connections. A gasket is compressed around the joint by the slip nut to form a watertight seal.

Slop sink   A deeper fixture than an ordinary sink, frequently installed in custodian's room.

Soil pipe   A pipe that conveys the discharge of water closets, or plumbing fixtures having similar functions, with or without discharges from other plumbing fixtures.

Soil stack   The main vertical stack that receives and conveys the discharge from all plumbing fixtures.

Solder   A metal alloy composed of tin and lead and used to join copper pipe and fittings.

Soldering iron   A tool composed of copper that is heated in a furnace and used to melt solder when joining pieces of metal.

Solder joint   The means of joining copper pipe to slip on fittings using solder.

Specifications   A document that describes the quality of materials and the work quality required for a given building. Specifications are the plumber's source of information about the quality of the pipe, fixtures, and so on, to be included in the plumbing system.

Spigot   The plain end of a cast-iron pipe. The spigot is inserted into the bell end of the next pipe to make a watertight joint.

Splash guard   A specially formed block that is placed under the outlet of a downspout to prevent erosion of the soil.

Spout   The end of a faucet that serves as a passageway for water out of the piping system.

Stack   A general term for the vertical main of a system of soil, waste, or vent piping.

Stack clean-out   A plugged fitting located at the base of all soil or waste stacks.

Stack vent   The extension of a soil or waste stack above the highest connected horizontal branch.

Star drill   A tool made from steel that has a star-shaped chisel on one end and a face that is hit with a hammer on the other end. This tool is used to make holes in concrete and masonry block.

Stem   The shaft of a faucet that holds the washer and to which the handle is attached.

Stock or die stock   A tool used to turn a die when cutting threads.

Stop and waste valve   A gate or compression type of valve that has a side opening, or port, and may be opened to allow water to drain from the piping supplied by the valve.

Stop box or curb box   An adjustable cast-iron box that is brought up to grade with a removable iron cover.

Stopper   A plug that controls wastewater drainage from a lavatory or bathtub, usually controlled remotely by a handle on the fixture; sometimes called a pop-up plug.

Storm drain   A drain that conveys rainwater, subsurface water, or other waste that does not need to be treated in a private or public sewage treatment facility.

Storm sewer   A sewer designed to convey only surface or storm water.

Storm water   The excess rainfall that runs off during or after a rain.

Storm water drainage system   The piping system used for conveying rainwater or other precipitation to the storm sewer or other place of disposal.

S-trap   An S-shaped, water-sealed trap sometimes used in plumbing. (Most water closet traps are S-traps.)

Strap wrench   A tool for gripping pipe. (The strap is made of nylon web treated with latex.)

Street L   An elbow fitting with one male end and one female end; it is the same as a service L.

Street T   A T with one female and one male threaded opening, plus an outlet opening with female threads.

Subfloor   A rough floor consisting of boards or plywood panels applied directly over the floor joist.

Subsoil drain   A drain that receives only subsurface water and conveys it to a storm drain.

Sump   A tank or pit installed in the basement of a building to collect subsurface water so it can be pumped to a storm drain.

Sump pump A rotary type of pump that lifts water from the sump into a drain pipe.

Supports, hangers, anchors Devices for securing pipes to walls, ceilings, floors, or other structural members, and plumbing fixtures to floors or walls.

Survey A description of a piece of property, including the measurements and marking of land.

Swage To increase or decrease the diameter of a pipe by using a special tool that is forced into or around the pipe.

Sweat soldering A method of soldering in which the parts to be joined are first coated with a thin layer of solder and then joined while exposed to a flame.

Swing joint A joint in a threaded pipe line that permits fittings to be installed in a close space.

Tamp To firmly compact earth during backfilling.

Tap A tool rotated by hand or machine to produce internal threads.

Tapered reamer A tool for deburring and cleaning the inside ends of pipes.

Tapped T A cast-iron T with at least one branch tapped to receive a threaded pipe or fitting.

Temperature and pressure relief valve A safety valve designed to protect against dangerous conditions by relieving high temperature and/or high pressure from a water heater.

Thermostat An automatic device consisting of a temperature-sensing unit that turns an energy source on and off; it is used in heating and cooling.

Three-quarter bath A bathtub containing a water closet, a lavatory, and a shower.

Three-quarter S-trap A trap shaped like three-fourths of the letter S.

Trap A fitting or device designed and constructed to provide, when properly vented, a liquid seal that prevents the passage of air without materially affecting the flow of liquid through it.

Trap arm   That portion of a plumbing fixture drain between the trap weir and the vent pipe connection.

Trap seal   The vertical distance between the crown weir and the dip of a trap that determines the depth of the water seal of a trap.

Trim   The water supply and drainage fittings that are installed on the fixture to control the flow of water into the fixture and the flow of wastewater from the fixture to the sanitary drainage system.

Trunk line   The main piping from which building drains and/or water supply piping branch off.

Tube   A conduit or conductor of cylindrical shape, the wall thickness of which is less than that needed to receive a standard pipe thread.

Tubing   Any thin-walled pipe that can be bent easily.

Tubing cutter   A tool used to cut tubing.

Underground piping   Piping in contact and covered with earth.

Union   A fitting used to join two lengths of pipe to permit disconnecting of the two without cutting them.

Unit vent   One vent pipe that serves two or more traps.

Vacuum breaker   A device that prevents the formation of a vacuum in a water supply pipe; it is installed to prevent back-flow.

Valve   A fitting installed by plumbers on a piping system to control the flow of fluid within that system.

Vanity   A bathroom fixture consisting of a lavatory set into or onto the top of a cupboard or cabinet.

Vapor barrier   A material that prevents moisture from penetrating a wall, ceiling, or floor.

Vent   That part of the drain, waste, or vent piping that permits air to circulate and protects the seals in traps from siphonage and back-pressure.

Vent, circuit   A vent installed where two similar fixtures discharge into a horizontal waste branch.

Vent, common   A vent that serves two or more fixture traps.

Venting, individual   The venting of each trap.

Vent, looped   A vent that drops below the flood rim of a fixture before being connected to the main vent.

Vent, pipe   The pipe installed to ventilate a building drainage system and to prevent trap siphonage and back-pressure.

Vent, relief   A vent installed at a point where the waste piping changes direction.

Vent stack   The vertical portion of the vent piping that extends through the roof of the building.

Vent system   A pipe or pipes installed to provide a flow of air to or from a drainage system or to provide a circulation of air within such a system in order to protect trap seals from siphonage and back pressure.

Vent, wet   A pipe that serves as both a vent and a drain.

Wall hung   Referring to a plumbing fixture supported by the wall.

Waste   A liquid discharged from a fixture; the liquid contains no fecal matter.

Waste pipe   A pipe that conveys liquid waste that does not contain fecal matter.

Water closet   A toilet.

Water conditioner   A device used to remove dissolved minerals from water.

Water cooler   An electric appliance that combines a drinking fountain with a water cooling unit.

Water distributing pipe   The piping that conveys water from a water service pipe to the fixture branch.

Water distributing system   The piping that conveys water from a service pipe to plumbing fixtures on another outlet.

Water hammer   The banging noise in pipes caused by a fast-closing valve.

Water heater An appliance for heating and supplying the hot water used within a building for purposes other than space heating.

Water main A large water supply pipe, generally located near the street, that serves a large number of buildings.

Water meter A device used to measure the amount of water in cubic feet or gallons that passes through the water service.

Water outlets A discharge opening in a water supply system of a building or premises through which water can be obtained for the several purposes for which it is used by means of a faucet, valve, or other control mechanism.

Water service pipe The pipe conveying water from a water main or other source of water supply to the water distributing system of a building.

Water softener An appliance that removes dissolved minerals (calcium and magnesium) from water by the process of ion exchange.

Water supply fixture unit (WSFU) A common measure of the probable hydraulic demand on the water supply by various types of plumbing fixtures.

Water supply system The water service pipe, water distributing system, fittings, and their accessories in or adjacent to any building, structure, or conveyance.

Wet vent A soil or waste pipe serving as a vent.

Working drawings Drawings showing exactly how a building should be constructed.

Yarning iron A tool used to pack oakum into bell-and-spigot pipe joints before they are leaded.

Yoke vent A vent pipe installed from a soil or waste stack that connects to a vent stack at a higher elevation for the purpose of preventing pressure changes in the two stacks.

Y (or Wye) branch A section of pipe that joins the main run of pipe at an angle. The fitting that makes the joint is in the shape of the letter Y.

22

Odds and Ends

If you feel as though something is missing from this book, you will probably find it in this chapter. You know that box that you keep on your truck that is full of unusual fittings and devices that you never know when you will need? Well, this chapter is to this book what that box is to your truck. As you peruse these pages, you will discover a lot of little facts that you may have never thought of. Although you probably won't use the information every day, it will be there when you need it.

COMMONLY USED ABBREVIATIONS

ABS	acrylonitrile butadiene styrene
AGA	American Gas Association
AWWA	American Water Works Association
BOCA	Building Officials Conference of America
B&S	bell-and-spigot (cast-iron) pipe
BT	bathtub
C-to-C	center-to-center
CI	cast iron
CISP	cast-iron soil pipe
CISPI	Cast Iron Soil Pipe Institute
CO	clean-out
CPVC	chlorinated polyvinyl chloride
CW	cold water
DF	drinking fountain
DWG	drawing
DWV	drainage, waste, and vent system
EWC	electric water cooler
FG	finish grade
FPT	female pipe thread
FS	federal specifications
FTG	fitting
FU	fixture-unit
GALV	galvanized
GPD	gallons per day
GPM	gallons per minute
HWH	hot-water heater
ID	inside diameter
IPS	iron pipe size
KS	kitchen sink
LAV	lavatory
LT	laundry tray
MAX	maximum
MCA	Mechanical Contractors Association
MGD	million gallons per day
MI	malleable iron
MIN (min)	minute or minimum

COMMONLY USED ABBREVIATIONS *(cont.)*

MPT	male pipe thread
MS	mild steel
M TYPE	lightest type of rigid copper pipe
NAPHCC	National Association of Plumbing, Heating, and Cooling Contractors
NBFU	National Board of Fire Underwriters
NBS	National Bureau of Standards
NPS	nominal pipe size
NFPA	National Fire Protection Association
OC	on center
OD	outside diameter
SAN	sanitary
SHWR	shower
SV	service
S & W	soil & waste
SS	service sink
STD (std)	standard
VAN	vanity
VTR	vent through roof
W	waste
WC	water closet
WH	wall hydrant
WM	washing machine
XH	extra heavy

AVAILABLE LENGTHS OF COPPER PLUMBING TUBE

Drawn (hard copper) (feet)		*Annealed (soft copper) (feet)*	
Type K Tube			
Straight Lengths:		Straight Lengths:	
Up to 8-in. diameter	20	Up to 8-in. diameter	20
10-in. diameter	18	10-in. diameter	18
12-in. diameter	12	12-in. diameter	12
		Coils:	
		Up to 1-in. diameter	60
			100
		1¼-in. diameter	60
			100
		2-in. diameter	40
			45
Type L Tube			
Straight Lengths:		Straight Lengths:	
Up to 10-in. diameter	20	Up to 10-in. diameter	20
12-in. diameter	18	12-inch diameter	18
		Coils:	
		Up to 1-in. diameter	60
		100	
		1¼- and 1½-in. diameter	60
		100	
		2-in. diameter	40
		45	
DWV Tube			
Straight Lengths:		Not available	
All diameters	20		
Type M Tube			
Straight Lengths:		Not available	
All diameters	20		

COPPER TUBE

Inside Diameter (inches)	Nominal Size (inches)	Outside Diameter (inches)
Type DWV		
N/A	1/4	0.375
N/A	3/8	0.500
N/A	1/2	0.625
N/A	5/8	0.750
N/A	3/4	0.875
N/A	1	1.125
1.295	1¼	1.375
1.511	1½	1.625
2.041	2	2.125
3.035	2½	2.625
N/A	3	3.125
N/A	3½	3.625
4.009	4	4.125
4.981	5	5.125
5.959	6	6.125
N/A	8	8.125
N/A	10	10.125
N/A	12	12.125

COPPER TUBE *(cont.)*

Nominal Pipe Size (inches)	Outside Diameter (inches)	Inside Diameter (inches)
Type K		
1/4	0.375	0.305
3/8	0.500	0.402
1/2	0.625	0.527
5/8	0.750	0.652
3/4	0.875	0.745
1	1.125	0.995
1¼	1.375	1.245
1½	1.625	1.481
2	2.125	1.959
2½	2.625	2.435
3	3.125	2.907
3½	3.625	3.385
4	4.125	3.857
5	5.125	4.805
6	6.125	5.741
8	8.125	7.583
10	10.125	9.449
12	12.125	11.315
Type L		
1/4	0.375	0.315
3/8	0.500	0.430
1/2	0.625	0.545
5/8	0.750	0.666
3/4	0.875	0.785
1	1.125	1.025
1¼	1.375	1.265
1½	1.625	1.505
2	2.125	1.985
2½	2.625	2.465
3	3.125	2.945
3½	3.625	3.425
4	4.125	3.905
5	5.125	4.875
6	6.125	5.845
8	8.125	7.725
10	10.125	9.625
12	12.125	11.565

COPPER TUBE *(cont.)*

Nominal Pipe Size (inches)	Outside Diameter (inches)	Inside Diameter (inches)
Type M		
1/4	0.375	0.325
3/8	0.500	0.450
1/2	0.625	0.569
5/8	0.750	0.690
3/4	0.875	0.811
1	1.125	1.055
1¼	1.375	1.291
1½	1.625	1.527
2	2.125	2.009
2½	2.625	2.495
3	3.125	2.981
3½	3.625	3.459
4	4.125	3.935
5	5.125	4.907
6	6.125	5.881
8	8.125	7.785
10	10.125	9.701
12	12.125	11.617

POLYVINYL CHLORIDE PLASTIC PIPE (PVC)

Nominal Pipe Size (inches)	Outside Diameter (inches)	Inside Diameter (inches)	Wall Thickness (inches)
1/2	0.840	0.750	0.045
3/4	1.050	0.940	0.055
1	1.315	1.195	0.060
1¼	1.660	1.520	0.070
1½	1.900	1.740	0.080
2	2.375	2.175	0.100
2½	2.875	2.635	0.120
3	3.500	3.220	0.140
4	4.500	4.110	0.195

BRASS PIPE

Nominal Pipe Size (inches)	Outside Diameter (inches)	Inside Diameter (inches)	Wall Thickness (inches)
1/8	0.405	0.281	0.062
1/4	0.376	0.376	0.082
3/8	0.675	0.495	0.090
1/2	0.840	0.626	0.107
3/4	1.050	0.822	0.144
1	1.315	1.063	0.126
1¼	1.660	1.368	0.146
1½	1.900	1.600	0.150
2	2.375	2.063	0.156
2½	2.875	2.501	0.187
3	3.500	3.062	0.219

STEAM PIPE EXPANSION
(INCHES INCREASE PER 100 IN.)

Temperature (°F)	Steel	Cast Iron	Brass and Copper
0	0	0	0
20	0.15	0.10	0.25
40	0.30	0.25	0.45
60	0.45	0.40	0.65
80	0.60	0.55	0.90
100	0.75	0.70	1.15
120	0.90	0.85	1.40
140	1.10	1.00	1.65
160	1.25	1.15	1.90
180	1.45	1.30	2.15
200	1.60	1.50	2.40
220	1.80	1.65	2.65
240	2.00	1.80	2.90
260	2.15	1.95	3.15
280	2.35	2.15	3.45
300	2.50	2.35	3.75
320	2.70	2.50	4.05
340	2.90	2.70	4.35
360	3.05	2.90	4.65
380	3.25	3.10	4.95
400	3.45	3.30	5.25
420	3.70	3.50	5.60
440	3.95	3.75	5.95
460	4.20	4.00	6.30
480	4.45	4.25	6.65
500	4.70	4.45	7.05

BOILING POINT OF WATER AT VARIOUS PRESSURES

Vacuum in Inches of Mercury	Boiling Point	Vacuum in Inches of Mercury	Boiling Point
29	76.62	14	181.82
28	99.93	13	184.61
27	114.22	12	187.21
26	124.77	11	189.75
25	133.22	10	192.19
24	140.31	9	194.50
23	146.45	8	196.73
22	151.87	7	198.87
21	156.75	6	200.96
20	161.19	5	202.25
19	165.24	4	204.85
18	169.00	3	206.70
17	172.51	2	208.50
16	175.80	1	210.25
15	178.91		

WEIGHT OF CAST-IRON SOIL PIPE

Size (inches)	Service Weight Per Linear Foot (pounds)	Extra Heavy Size (inches)	Per Linear Foot (pounds)
2	4	2	5
3	6	3	9
4	9	4	12
5	12	5	15
6	15	6	19
7	20	8	30
8	25	10	43
		12	54
		15	75

SIZES OF ORANGEBURG[a] PIPE

Inside Diameter (inches)	Length (feet)
2	5
3	8
4	8
5	5
6	5

[a]A cardboard-type drain pipe.

WEIGHT OF CAST-IRON PIPE

	Diameter (inches)	Service Weight (lb)	Extra Heavy Weight (lb)
Double Hub, 5-ft Lengths	2	21	26
	3	31	47
	4	42	63
	5	54	78
	6	68	100
	8	105	157
	10	150	225
Double Hub, 30-ft Length	2	11	14
	3	17	26
	4	23	33
Single Hub, 5-ft Lengths	2	20	25
	3	30	45
	4	40	60
	5	52	75
	6	65	95
	8	100	150
	10	145	215
Single Hub, 10-ft Lengths	2	38	43
	3	56	83
	4	75	108
	5	98	133
	6	124	160
	8	185	265
	10	270	400
No-Hub Pipe, 10-ft Lengths	1½	27	
	2	38	
	3	54	
	4	74	
	5	95	
	6	118	
	8	180	